# INTRODUCTION
## TO
# STATISTICAL METHODS
## IN
# MODERN GENETICS

## Asian Mathematics Series

A series edited by Chung-Chun Yang, *Department of Mathematics*
*The Hong Kong University of Science and Technology, Hong Kong*

**Volume 1**
Dynamics of Transcendental Functions
*Xin-Hou Hua and Chung-Chun Yang*

**Volume 2**
Approximate Methods and Numerical Analysis for Elliptic Complex Equations
*Goa-Chun Wen*

**Volume 3**
Introduction to Statistical Methods in Modern Genetics
*Mark C. K. Yang*

## In Preparation

Mathematical Theory in Periodic Plane Elasticity
*Jian-Ke Lu and Hai-tao Cai*

# INTRODUCTION
## TO
# STATISTICAL METHODS
## IN
# MODERN GENETICS

Mark C. K. Yang

Department of Statistics
University of Florida, USA

**CRC Press**
Taylor & Francis Group
Boca Raton London New York

CRC Press is an imprint of the
Taylor & Francis Group, an **informa** business

CRC Press
Taylor & Francis Group
6000 Broken Sound Parkway NW, Suite 300
Boca Raton, FL 33487-2742

First issued in paperback 2019

ISBN-13: 978-90-5699-134-0 (hbk)
ISBN-13: 978-0-367-39890-3 (pbk)

### Library of Congress Cataloging-in-Publication Data

Yang, Mark C. K.
  Introduction to statistical methods in modern genetics. —
  (Asian mathematics series; v.3)
        p.  cm.
  Includes bibliographical references and index.
  ISBN 90-5699-134-5    ISSN: 1028-1428
  1.Genetics—Statistical methods.
  I. Title.  II. Yang, Zhongjun
576.5′015195

**Visit the Taylor & Francis Web site at**
**http://www.taylorandfrancis.com**

**and the CRC Press Web site at**
**http://www.crcpress.com**

# Contents

# Introduction to the Series

The *Asian Mathematics Series* provides a forum to promote and reflect timely mathematical research and development from the Asian region, and to provide suitable and pertinent reference or text books for researchers, academics and graduate students in Asian universities and research institutes, as well as in the West. With the growing strength of Asian economic, scientific and technological development, there is a need more than ever before for teaching and research materials written by leading Asian researchers, or those who have worked in or visited the Asian region, particularly tailored to meet the growing demands of students and researchers in that region. Many leading mathematicians in Asia were themselves trained in the West, and their experience with Western methods will make these books suitable not only for an Asian audience but also for the international mathematics community.

The *Asian Mathematics Series* is founded with the aim to present significant contributions from mathematicians, written with an Asian audience in mind, to the mathematics community. The series will cover all mathematical fields and their applications, with volumes contributed to by international experts who have taught or performed research in Asia. The level of the material will be at graduate level or above. The book series will consist mainly of monographs and lecture notes but conference proceedings of meetings and workshops held in the Asian region will also be considered.

# Preface

The importance of genetics can be felt almost daily. Genetics news not only appears in scientific magazines such as *Science, Nature, Discovery* and *Scientific American* where genetics occupies a very large proportion, but also in daily newspapers where news of genes, cloning and genetically engineered new species attracts much public attention. Though I learned some statistics with genetic applications in my graduate study, they seemed to be very far from the excitement of present day genetics. I began to read papers in modern genetics to see what new statistical tools are used. I also attended several short courses and did a sequence of experiments in DNA recombination and protein manipulation. Out of the knowledge I gathered and the data analyses I performed for the medical school, I taught a genetic data analysis course several times. This book grew out of my lecture notes for the course.

There are several recent books on statistical methods in genetics, such as *Genetic Data Analysis II* by Bruce Weir, *Analysis of Human Genetic Linkage* by Jurg Ott and *Mathematical and Statistical Methods for Genetic Analysis* by Kenneth Lange. All are excellent books and I have used them in my class. Though there is some overlap between this book and theirs, the overlap is not very much. My emphasis is on understanding how and why statistical concepts and methods are needed and used.

There are numerous topics that one can select in modern genetics. The topics I chose are undoubtedly biased, but these were the topics I wanted to know more about when I got into this field. I feel that many beginners will share my interest in these topics. They include:

- How is a gene found?
- How have scientists separated the genetic and environmental aspects of a person's intelligence?
- How have genetics been used in agriculture so that the domestic animals and crops are constantly improving?
- What is a DNA fingerprint and why are there controversies about it?
- How did they use genes to rebuild the evolutionary history?

I endeavored to understand them and I believe I do now. I hope that in this book you will find no gaps in how these questions are answered. Though the statistical theory behind basic modern genetics is not difficult, most statistical genetics papers are not easy to read for a beginner. Formulae quickly become very tedious to fit a particular application of genetics. This book tries to distinguish between the necessary and unnecessary complexity in presenting the material to a statistician. The details for specific applications are referred to as references.

Since this book is written mainly for statistics students, a certain statistical background is assumed. The necessary background statistical knowledge beyond elementary statistics is given in Appendix A. One can easily see whether one has enough background for this book. One year of graduate study in most statistics departments should be sufficient.

# Notations

| | |
|---|---|
| A, a; B, b; D, d | alleles |
| Cov() | covariance |
| E() | expectation of a random variable |
| F | pedigree information |
| $f$ | frequency |
| $f()$ | function, probability density function |
| g | genotype |
| $h^2$ | heritability |
| IBD | identity by descent |
| M, m | marker information |
| mle | maximum likelihood estimate |
| p( ), P( ), $Pr\{\}$ | probability |
| $P(A\|B)$, $Pr\{A\|B\}$ | conditional probability of $A$ given $B$ |
| $p$ | allele frequency |
| $q$ | allele frequency |
| Q | location of a quantitative gene |
| r | recombination fraction, sample correlation |
| S | alleles shared by two individuals |
| T | a test statistic |
| V(), Var() | variance |
| Z, z | LOD score |
| $z_\alpha$ | the upper $\alpha 100\%$ quantile value of a standard normal distribution |
| $\alpha$ | a parameter, or the significance level in hypothesis testing |
| $\beta$ | a parameter, power |
| $\delta^2$ | noncentrality parameter |
| $\varepsilon$ | error random variable |
| $\phi( . )$ | normal density, kinship coefficient |
| $\Phi$ | normal distribution function, kinship matrix, IBD sharing vector |
| $\lambda$ | affection rate, likelihood ratio |
| $\mu$ | mean |
| $\pi$ | the usual $\pi = 3.1416$, IBD allele shared by two individuals |
| $\rho$ | correlation coefficient |
| $\theta$ | recombination fraction, collection of parameters |
| $\sigma^2$ | variance |
| $\Psi$ | $\theta^2 + (1 - \theta)^2$, see Table 2.3.10 |
| $\Sigma$ | covariance matrix |
| $\equiv$ | defined as |
| $\square$ | end of a proof or an exampl᷉ |

# Chapter 1

# Background to Modern Statistical Genetics

## 1.1  Introduction: Where is the Gene?

One of the main goals in modern genetics is to locate the gene (or genes) that is (are) responsible for a certain trait. Once the gene is located, we can pursue the following investigations:

1. Identify the difference between a normal gene and an abnormal gene.
2. Find the protein (enzyme) that this gene produces and consequently treat patients with the normal protein.
3. Use the abnormal protein to understand different physiological pathways. For example, if a person cannot digest lactose because of a gene, then we know that the protein produced by this gene is in the pathway of lactose digestion.
4. By screening early, a person who is susceptible to a certain disease can be informed. Preventive measures may then be possible.
5. Once the difference between a bacteria gene and human gene that perform the same function is found, a drug that can kill the bacteria, but is not poisonous to humans, can be manufactured. The same idea can be used to treat cancer cells.
6. We may be able to remove or implant a gene in animals and plants so that the modified organism can express more desirable traits, such as frost resistance.

Let us start with human beings. Listed below are some basic genetic facts that one has probably learned from a biology course.

1. The genetic material is stored in the chromosomes in the nucleus of every cell.
2. Every normal person has 22 pairs (**diploids**) of chromosomes and a pair of sex chromosomes with XX for females and XY for males.

1

3. For each pair of chromosomes, one band (called **haploid** or **monoploid**) is from the father and one is from the mother. The pairs are called sister chromosomes. A gene is located in a particular site, called a **locus**, on a chromosome. The two genes on the same locus of the two sister chromosomes may not be exactly the same. Genes with different structures on the same location are called **alleles**.

4. When the egg or sperm cells are formed in the parents, the cell undergoes a division process called **meiosis** (see §1.2 for details), when the genetic materials in the two sister chromosomes are mixed. In other words, during meiosis two sister chromosomes act like two strings mingling together to exchange materials. This process is explained in Fig. 1.1.1. An egg or sperm cell contains genetic material from one of the two haploids after recombination. The genetic material that comes from the same parent is called a **gamete**.

5. The **recombination fraction** $\theta$ between two loci is defined as the probability that before meiosis the two genes were originally on the same haploids (e.g., B and 1 in Fig. 1.1.1 (1)), but after meiosis are on two different haploid (see Fig. 1.1.1 (3) for B and 1). Obviously, the closer the two genes, the less chance that they will be separated during the meiosis process. In other words, the closer the two loci, the smaller the recombination fraction. When two genes are on two

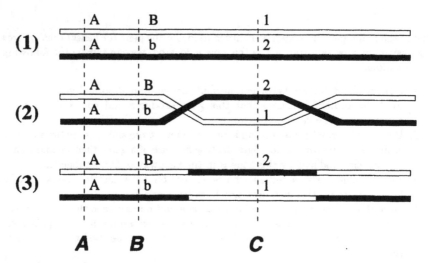

Figure 1.1.1: Basic symbols and terminologies in meiosis. (1) The two chromosomes or haploids in a parent's cell. The **A**, **B**, and **C** are loci of genes. The two alleles at the same locus are denoted by letters or numbers. For example, at locus $A$ the two alleles are $A$ and $A$, so it is a homozygous locus. However, at loci $B$ and $C$, they are heterozygous. (2) During meiosis, the two chromosomes may tangle together and exchange material. (Location and alleles sometimes share the same symbol, such as locus $B$ has alleles $B$ and $b$. It may cause some confusion, but it is a tradition in genetics.) (3) After meiosis, two gametes are formed. Each of them will go to one egg or sperm cell.

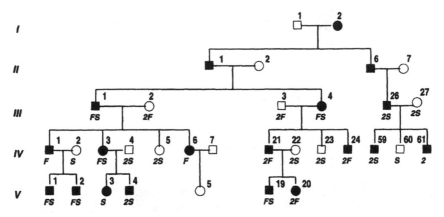

Figure 1.1.2: Pedigree for a *dentinogensis imperfecta* family. This is only part of the pedigree from Ball et al. (1982).

different chromosomes or are far apart on the same chromosome, $\theta = 1/2$. Thus, if the location of one gene, $A$, is known, then the position of the other gene, $B$, can be expressed by the recombination rate $\theta$ between $A$ and $B$. (The details are discussed in §1.2)

6. The parameter $\theta$ can be estimated by studying the family tree or the pedigree.

Fig. 1.1.2 shows a part of a pedigree of a family with *dentinogensis imperfecta*, a defect in dentin production giving teeth a brown or blue appearance (Ball et al., 1982). Some of the current family members have their GC protein typed. The GC serum protein, which is responsible for protecting hemoglobin in the urine filtration process, has three varieties, GC1S, GC1F and GC2 and is known to be controlled by one gene. Thus, GC1S, GC1F and GC2 can be considered as the alleles of the *GC* gene and they are marked as $S$, $F$ and 2 in the pedigree. The usual symbols in a pedigree are empty squares for unaffected males, dark squares for affected males, empty circles for unaffected females, and dark circles for affected females. From this pedigree, it is apparent that the disease trait must be **dominant**, i.e., if you get one disease allele in the diploid, you have the disease. The other type of gene is called **recessive**, i.e., a person will not have the disease unless the disease gene appears in both bands of the chromosome pair. We will learn more about this in §1.2.

The lines represent marriages and descendants in an obvious way. The Roman numerals on the left of the pedigree represent generations and the number beside each individual is used to identify that person. For example, individual III-1 is an affected male with GC protein alleles $F$ and $S$ and individual IV-2 is an unaffected female with GC composition $S$ only. When a person has two different alleles at the same locus, such as III-1, his **genotype** is called **heterozygous**. On the other hand, the genotype of IV-2 is called **homozygous**. In the original paper, Ball et al. (1982) were interested in finding the distance between the *dentinogensis imperfecta* gene and GC gene by estimating their recombination fraction $\theta$ from this pedigree.

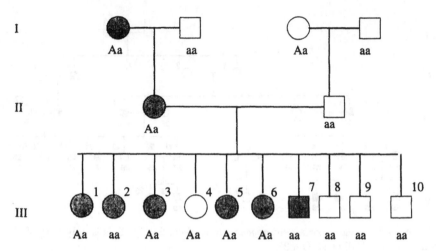

Figure 1.1.3: A simple pedigree for a dominant disease trait and alleles $(A, a)$.

Note that, in general, we cannot tell whether the disease gene and one of the GC alleles are on the same haploid (or gamete) or not. For example, for subject III-1, we only know that he is affected and his serum contains proteins GC1F and GC1S. Let $D$ denote the disease gene and $d$ denote the normal gene. Person III-1 must have $Dd$ genotype because one of his children is not affected. Whether the $D$ and $S$ are on the same haploid, denoted by $DS/dF$, or $D$ and $F$ are on the same haploid, denoted by $DF/dS$, cannot be determined by this individual alone. But we may know the **phase** information by studying a pedigree, as we shall see later in this section. The analysis of a large pedigree such as the one in Fig. 1.1.2 will be delayed until §2.2. Here, only the elementary concept is presented with a simple example shown in Fig. 1.1.3. Suppose the disease gene is dominant and $A$ and $a$ denote the two alleles of another gene with known location.

Suppose at this moment only the generation II and III data were available. Obviously, the phase of the affected mother is unknown. She may be $DA/da$ or $dA/Da$. Supposed she is $DA/da$ (she cannot have $DD$, otherwise all of her children would be affected.). Given the recombination fraction $\theta$, the likelihood of the ten offspring having 3 recombinations (individuals 2, 4 and 7) is

$$\binom{10}{3}\theta^3(1-\theta)^7. \tag{1.1.1}$$

However, there is the other possibility of her being $dA/Da$. Given no other information, the chance of her being $DA/da$ and $dA/Da$ is half and half. Thus the likelihood for $\theta$ based on Fig.1.1.3 is

$$L(\theta) = \frac{1}{2}\binom{10}{3}[\theta^3(1-\theta)^7 + \theta^7(1-\theta)^3]. \tag{1.1.2}$$

There is no analytical expression for the maximum likelihood estimation (mle) for $\theta$. Numerical solution gives the mle $\hat{\theta} = 0.318$. This means that the disease gene and the $A - a$ genes are not very closely **linked**. Note that two loci are not linked if $\theta = 1/2$.

Since the data from the first generation I are available in Fig. 1.1.3, the affected mother in generation II must be $DA/da$. Thus, the likelihood function for $\theta$ is (1.1.1), and the mle $\hat{\theta} = 0.3$. Though the new estimated value is similar to the mle using (1.1.2), the current one should be more accurate. We leave this as an exercise.

This is the basic idea in classic linkage analysis. One difficult part is that we have to know the position of one gene that is close to the gene we wish to find. Thus, in the early days researchers were particularly interested in the linkage relation between a characteristic and a few known genes. For example, Hogben and Pollack (1935) and Fisher (1936) tried to establish the linkage between Friedreich's ataxia and blood type. Zieve et al. (1936) and Finney (1940) were interested in the blood type gene and allergy. In modern genetics, many location identifiers, called **markers**, are available through chemical analysis of the chromosome. The rest of this chapter is devoted to the biological and chemical background of modern genetics. Two short expository articles (Yang, 1995a,b) also give an introduction to the applications of statistics in modern genetics.

## 1.2 Function of Genes in Reproduction

### 1.2.1 Mendel's Law of Heredity and Early Development in Genetics

Humans have long known that characteristics can be passed down through generations, but it was difficult to see how this was done. Though we say, "Like father like son", we have also observed that tall parents can have short children. Not too much progress in verifiable description of inheritance was made until Mendel conducted his experiments around 1865.

Gregor Mendel studied garden peas (*Pisum sativum*) in a series of experiments in his Brno monastery, Czechoslovia. The garden pea has a few varieties; yellow or green seed color, round or wrinkled seed skin, and 5 other characteristics (called **phenotypes**) listed in Table 1.2.1. Peas usually maintain the same characteristics from generation to generation because peas are self-pollinating plants. Mendel cross-pollinated plants by removing the pollen-bearing anthers from one plant and brushing its stigma with the pollen from another plant with the opposite trait. The offsprings are called the **first filial generation** ($F_1$). It was surprising that in Mendel's experiments all plants in the $F_1$ generation had the same characteristics. For example, when plants with purple and white flowers were cross-bred, no matter which pollen or which stigma was used, all $F_1$ had purple flowers. This was very unusual, because one might expect some purple, some white and some mixed color in the $F_1$ generation. However, in Mendel's experiments (listed in Table 1.2.1), one character was always dominant. He

Table 1.2.1: Results of Mendel's experiments on seven pea characters.

| Parent phenotypes | $F_1$ | $F_2$ | $F_2$ ratio |
|---|---|---|---|
| 1. Round × wrinkled seeds | All round | 5474 round; 1850 wrinkled | 2.96 : 1 |
| 2. Yellow × green seeds | All yellow | 6022 yellow; 2001 green | 3.01 : 1 |
| 3. Purple × white petals | All purple | 705 purple; 224 white | 3.15 : 1 |
| 4. Inflated × pinched pods | All inflated | 882 inflated; 299 pinched | 2.95 : 1 |
| 5. Green × yellow pods | All green | 428 green; 152 yellow | 2.82 : 1 |
| 6. Axial × terminal flowers | All axial | 651 axial; 207 terminal | 3.14 : 1 |
| 7. Long × short stems | All long | 787 long; 277 short | 2.84 : 1 |

then let the $F_1$ plants self-fertilize and found the recessive character appeared again and the ratio for the dominant trait to the recessive was always close to 3 : 1, (see Table 1.2.1 last column). From this, Mendel hypothesized his first two laws of inheritance.

**Mendel's first law:** An individual's character is determined by a pair of genetic material called genes (**alleles**). The two alleles separate (**segregate**) from each other during sex-cell (pollen or ovule) formation. Each sex-cell takes one of the two alleles with equal probability.

**Mendel's second law:** If the two alleles are different (**heterozygatic**) in an individual, then one will dominate the other in the phenotype.

Mendel used algebraic symbols for his theory (see Mendel, 1867). He used capital $A$ for the dominant trait and small letter $a$ for the recessive one. He explained the 3:1 ratio as shown in Fig. 1.2.1.

$$
\begin{array}{ccccc}
F_0 & & AA(\mathbf{A}) & \times & aa(\mathbf{a}) \\
& & & \downarrow & \\
F_1 & & & Aa(\mathbf{A}) & \\
& & & \downarrow & \\
F_2 & AA(\mathbf{A}) & Aa(\mathbf{A}) & & aA(\mathbf{A}) \quad aa(\mathbf{a})
\end{array}
$$

Figure 1.2.1: Model for Mendel's first two laws. The symbols are: genotype (phenotype). For example, the genotype of $F_1$ is $Aa$ and the phenotype is $\mathbf{A}$. $F_0$ are two pure-bred parents with two different phenotypes. $F_1$ are their offspring and $F_2$ are the offspring of $F_1$ by self-pollination (**intercross**).

We now know that Mendel's second law is not infallible because there are many **co-dominant** alleles, i.e., the expression of $Aa$ is not the same as $AA$ or $aa$. For example, cross-breeding of a pure red four-o'clock (a North American wild flower that resembles a morning glory) and a pure white one will produce pink $F_1$, and the self-fertilized $F_2$ will produce 1:2:1 red, pink and white flowers.

When Mendel studied two characteristics simultaneously, such as the color and the shape of the peas, he found that the results agreed with the independent assortment

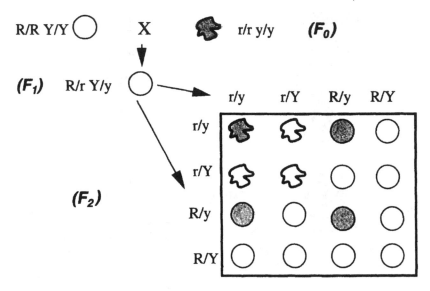

Figure 1.2.2: Mendel's third law (independent assortment): The $F_0$ generation are two pure-breds, one with R/R (round, dominant) Y/Y (yellow, dominant) and the other with r/r (wrinkled, recessive) y/y (green, recessive). In $F_1$ all the genotypes are R/r and Y/y with phenotype round and yellow. In $F_2$, by independent assortment of these two trait, gametes r/y, r/Y, R/y, and R/Y have equal probability to pass into the sperm or egg cell. Thus the phenotypes become, wrinkled/green : wrinkled/yellow : round/green : round/yellow $= 1 : 3 : 3 : 9$.

hypothesis, i.e., the two traits act independently in their segregation (see Fig. 1.2.2). This was called Mendel's third law. Again this law is not always true. Genes located together do not segregate independently. However, it was true in Mendel's case because all the seven traits he studied were on seven different chromosomes in peas and peas have only seven chromosomes. Thus, Mendel's work represented many key elements of great discoveries; right materials, well designed and carefully conducted experiments, a large number of data, sound mathematical modeling and luck.

Not only do genes close together not satisfy the independent assortment hypothesis, but two genes on different chromosomes may alter the ratios in Fig. 1.2.2 due to gene interactions, called **epistasis** in genetics. The albino gene has this property. Let $C$ be the dominant color expression allele and $c$ be the albino recessive allele which prevents color expression. Thus, any color gene, say $B$ (dominant) and $b$ (recessive), will be affected by this gene. For example, $BBcc$, $Bbcc$, and $bbcc$ all become albino, but $B$ and $b$ alleles express normally with $CC$ or $Cc$ genotypes. More examples can be found in Gardner et al. (1991). Another possible violation of Mendel's laws is *genome imprinting*: a genetic chemical that can temporarily and erasably suppress a gene's activity by one of the parents, i.e., the same gene from the mother or from the father may be expressed differently (see Sapienza, 1990).

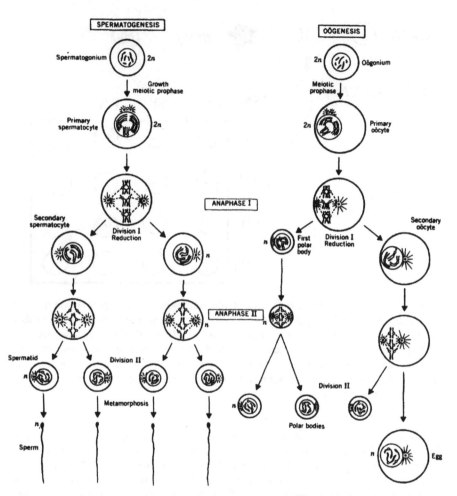

Figure 1.2.3: Meiotic sequence in male and female animals. The crossover happens at the end of anaphase I when one cell is divided into two cells. The detail is shown in Fig. 1.2.4. (From *Principles of Genetics*, by Gardner et al., copyright ©1991 by John Wiley & Sons, Inc. Reprinted by permission by John Wiley & Sons, Inc.)

Mendel's discoveries were published around 1865, but did not receive any significant response from other biologists until 1900 when biologists had already recorded the sequence of events during cell division under the microscope. They noticed a thread-like structure in cell nucleus which they named the chromosome. They also noticed the difference between body (**somatic**) cell division (called **mitosis**) and sex-cell division (called **meiosis**). Fig. 1.2.3 shows the detailed process of meiosis under a microscope. Apparently, a sex cell, such as the sperm or egg, has only half the number of chromosomes in the normal cell. In 1902, Walter Sutton and Theodor Boveri linked this phenomenon with Mendel's law and postulated that the genetic material must be in the chromosomes. Soon after them, Thomas Morgan found that the eye-color of the fruit fly must be linked to their sex-determining chromosome, which confirmed the Sutton-Boveri postulate.

The fruit fly (*Drosophila melanogaster*) is one of the most studied animals in genetics. Unlike peas, fruit flies grow quickly in the laboratory. Its maturation time from egg to adult takes only 12 to 14 days, and the animal with its 2 mm size occupies little space. Moreover, one single female can produce several hundred eggs and each has only four pairs of chromosomes.

One important step in meiosis is the crossover between the two haploids, each of which originated from one parent (Figs 1.1.1 and 1.2.4). It is during this process that sister chromosomes are mixed, and it is because of this process that large variations in offspring can be produced. Apparently, nature favors large variations in the

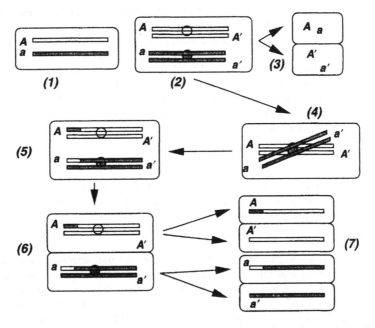

Figure 1.2.4: How the haploid chromsomes are transferred from the mother cell to the daughter cells during meiosis. Only one pair of chromosomes is used for demonstration. (1) Chromosomes *A* and *a* before meiosis. (2) At the beginning of prophase they are duplicated into *A* and *A'* and *a* and *a'*. The two haploids, one original and one duplicate, are connected in the center, called the centromere. (3) In normal cell division (mitosis) this cell is divided into two cells, each with one *A* and *a*. (4) In meiosis, crossover happens next. (5) After crossover, genetic material has been exchanged between *A* and *a* (Or *A'* and *a'*.) (6) The cell then divides. This division is called a "reduction" because each cell has one copy of the *A* or *a*. (7) The four gametes are separated into four sperm or egg cells.

reproduction process. Large variations introduce quick adaptation. It is difficult for a slow-adaptation species to compete with a fast-adaptation species.

## 1.2.2   Distance between Genes and Map Functions

In §1.1 we defined $\theta$ as the recombination probability between two loci in meiosis. Also, it was mentioned there that two genes far apart have a high chance of being separated during meiosis. Geneticists define the (linkage) distance between two gene as $\theta$ morgan (Morgan) when their recombinational fraction is $\theta$ ($\theta \leq 0.5$). For example, if $\theta = 0.02$ for genes A and B, we say that the distance between them is 0.02 morgan, or 2 centi-Morgan (cM). Fig. 1.2.5 shows the locations of various genes of *Drosophila*. For example, the dumpy-wing gene is located at 13 cM from the end of chromosome II, and the clot-eye gene is 16.5 cM from the end. Thus, the distance between these two genes is $16.5 - 13 = 3.5$ cM. Therefore, the chance that they are segregated during meiosis is approximately 0.035. Note that using probability to represent distance has its limitations. Distance can be extended indefinitely but probability cannot be larger than 1.0. Actually, the largest recombinational fraction between any two genes is 0.5 in genetics, but chromosome II in Fig. 1.2.5 is 107 cM long. In genetics, $x$ is usually used for the *physical distance* between two genes and $\theta$ is used for their crossover probability. The same unit, morgan, is used for both $x$ and $\theta$. The function that states their relationship is called a **map function**. Here we state only the most commonly used Haldane map function

$$\theta(x) = \frac{(1 - e^{-2|x|})}{2}. \tag{1.2.1}$$

For example, if two genes are 0.5 morgan away, then their recombination fraction is $\theta = [1 - \exp(-2 \times 0.5)]/2 = 0.316$. It can be easily seen that for a small $x$, $\theta(x) \approx x$. Some desirable properties of Haldane's map function are left in Exercise 1.7. More discussions of map function can be found in Ott (1991) and Liberman and Karlin (1984).

   Fig. 1.2.6 shows a schematic representation of human chromosomes and the numbering system. Counting chromosomes under the microscope was not an easy task. It was not until the late 1950s that we could confidently say that humans have 22 pairs of non-sex-related chromosomes (called **autosomes**) plus 1 pair of sex-linked chromosomes X and Y. A person with XX is a female and XY is a male. Because of this difference, any recessive gene on X becomes dominant in males. Color blindness and hemophila are two famous examples of sex-linked diseases that are predominant in males. There are now more than 200 traits identified as sex-linked. This partially explains why women on average live longer than men. As shown in Fig. 1.2.6, each chromosome consists of two arms; the short arm of a chromosome is indicated as $p$ and the long one as $q$. Each arm is further divided into regions and bands. For example, 9q34 is the gene at band 4 of region 3 on the long arm of chromosome 9. (This 9q34 is the region of the ABO blood type gene). It is estimated that human beings have 50,000 to 100,000 genes. The lengths of human chromosomes are given in Table 1.2.2 (Morton, 1991).

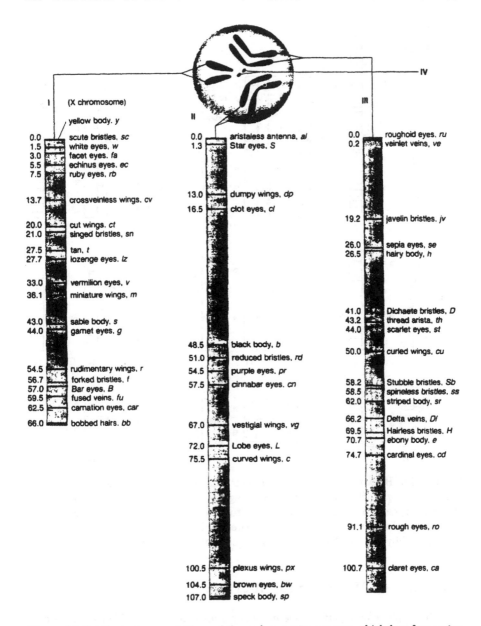

Figure 1.2.5: Physical map of *Drosophila melanogaster* genome which has four pairs of chromosomes. The number on the left is the distance in cM from the top of the chromosome to the gene whose function is described on the right side. (From *Genetics*, by Weaver and Hedrick (1989), copyright ©by The McGraw-Hill Companies. Reprinted by permission.)

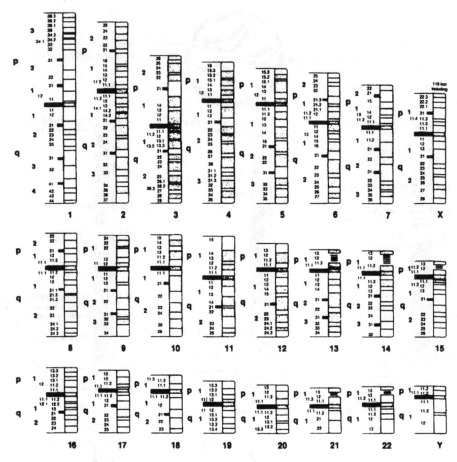

Figure 1.2.6: Human chromosomes and their numbering system. The two arms are called $p$ and $q$. Each arm is divided into regions and then into bands. The stripes (bands) represent how chromosomes react to different coloring chemicals for better visibility under a microscope. (From *Genetics*, by Weaver and Hedrick (1989), copyright ©by The McGraw-Hill Companies. Reprinted by permission.)

## 1.2.3  Hardy-Weinberg Equilibrium

Probability plays an important role in testing genetic hypotheses. Unlike the laws of physics and chemistry, most laws of genetics are subject to randomness and cannot be confirmed with a 100 percent predictable outcome. One important consequence of Mendel's laws is the possibility of Hardy-Weinberg Equilibrium (HWE) in a random mating population. Suppose there is one dominant and one recessive allele for a certain trait. An intuitive question would be "Will the dominant characteristic eventually

Table 1.2.2: Approximate lengths of human chromosomes measured in cM and in Mb. The meaning of Mb ($10^6$ bases) will be explained in §1.3.

| no. | 1 | 2 | 3 | 4 | 5 | 6 | 7 | 8 | |
|-----|-----|-----|-----|-----|-----|-----|-----|-----|---|
| length (Mb) | 236 | 255 | 214 | 203 | 194 | 183 | 171 | 155 | |
| length (cM) | 293 | 277 | 233 | 212 | 198 | 201 | 184 | 166 | |

| no. | 9 | 10 | 11 | 12 | 13 | 14 | 15 | 16 | |
|-----|-----|-----|-----|-----|-----|-----|-----|-----|---|
| length (Mb) | 145 | 144 | 144 | 143 | 114 | 109 | 106 | 98 | |
| length (cM) | 167 | 182 | 156 | 169 | 118 | 129 | 110 | 131 | |

| no. | 17 | 18 | 19 | 20 | 21 | 22 | X | Y | Total (with Y) |
|-----|-----|-----|-----|-----|-----|-----|-----|-----|----------------|
| length (Mb) | 92 | 85 | 67 | 72 | 50 | 56 | 164 | 59 | 3200 |
| length (cM) | 129 | 124 | 110 | 97 | 60 | 58 | 198 | – | 3702 |

dominate the whole population?" This seems likely because in Mendel's experiments there were more dominant traits expressed in both $F_1$ and $F_2$ generations. G. H. Hardy (1877–1947, a English mathematician) and W. Weinberg (1862–1937, a German physician) discovered independently within weeks that dominance will not result from random mating.

Suppose a population is large and there is (1) no immigration and emigration, (2) no cross generation breeding, (3) no mutation, and (4) no natural selection in favor of a certain trait. If at the 0th generation, the proportions (genotypic frequencies) of $AA$ $Aa$ and $aa$ are $u$, $v$, and $w$, respectively, with $u + v + w = 1$, then the genotypes of the next generation from random mating should be the results shown in Table 1.2.3.

Thus the expected proportions of the three traits in $F_1$ become:

$$p_1(AA) = u^2 + uv + v^2/4$$
$$p_1(Aa) = 2uw + vw + uv + v^2/2$$
$$p_1(aa) = w^2 + wv + v^2/4.$$

Table 1.2.3: Frequency relation by random mating between generations.

| Parents ($F_0$) | Frequency | Genotype of offsprings ($F_1$) |
|-----------------|-----------|-------------------------------|
| $AA \times AA$ | $u^2$ | $AA$ |
| $AA \times Aa$ | $2uv$ | $(1/2)AA + (1/2)Aa$ |
| $AA \times aa$ | $2uw$ | $Aa$ |
| $Aa \times Aa$ | $v^2$ | $(1/4)AA + (1/2)Aa + (1/4)aa$ |
| $Aa \times aa$ | $2vw$ | $(1/2)Aa + (1/2)aa$ |
| $aa \times aa$ | $w^2$ | $aa$ |

Note that the proportions may have changed, e.g., $p_1(AA)$ may not be the same as $p_0(AA) = u$. If we let $p = u + v/2$ and $q = 1 - p = w + v/2$, then

$$u_1 \equiv p_1(AA) = p^2, \quad v_1 \equiv p_1(Aa) = 2pq, \quad w_1 \equiv p_1(aa) = q^2.$$

It is surprising to find that generation $F_2$ resulting from the random mating of $F_1$ gives:

$$\begin{aligned} p_2(AA) &= u_1^2 + u_1 v_1 + v_1^2/4 \\ &= p^4 + 2p^3 q + p^2 q^2 \\ &= p^2, \end{aligned}$$

and $p_2(Aa) = 2pq$ and $p_2(aa) = q^2$. The proportions of the phenotypes reach equilibrium after one generation.

Apparently the assumption that non-cross generation breeding is no longer needed for the HWE after the first generation, and if the same proportions of each genotypes emigrate out, the equilibrium is not affected. The assumption that there is no natural selection may not be realistic. However, natural selection, in most cases, is a very slow process, but HWE can be reached in one generation. A large population is needed for the true proportions to equal the expected proportions. If the population is small in every generation, there will be random drift in the gene composition. We will see this in §4.3.

## 1.3   Some Facts in Molecular Genetics

The next two sections of this chapter present a brief introduction to molecular genetics. This knowledge is essential to understand the statistical methods in modern genetics. Acronyms such as RFLP, PCR and VNTR often appear in statistical methodological papers. The best way to understand these terms is to do the related experiments. Next to doing the experiments is to understand the facts and logic behind the experimental procedures so that we are confident that the assumptions behind the statistical methods make sense.

Fig. 1.3.1 explains the structure of a chromosome. Starting from the top left of the figure, a chromosome is unwound step by step until it reaches its most detailed level, the chemical structure. At the extreme right of the figure is the famous double-helix model suggested by Watson and Crick in 1955. This model was discovered by using X-ray diffraction on the basic compound in chromosome, the deoxyribonucleic acid (**DNA**) molecule. It is beyond the scope of this book to discuss many of the interesting experiments that led to the discovery of the DNA structure. Only some important facts will be stated using Fig. 1.3.1 as a basis.

(1) The genetic information is stored using four genetic codes called A (adenine), T (thymine), G (guanine) and C (cytosine). The structures of A and G are similar.

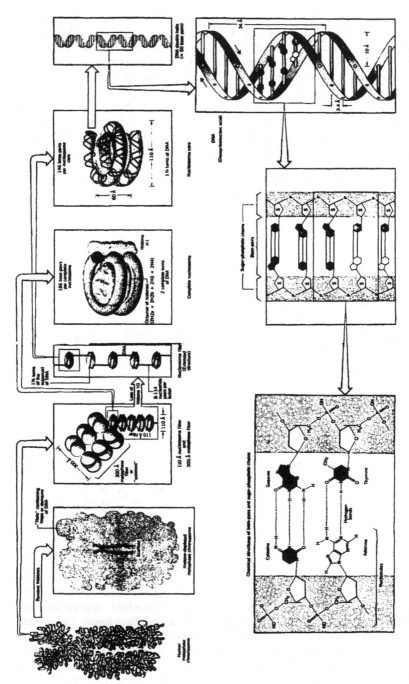

Figure 1.3.1: From chromosome to atomic structure of DNA. (From *Principles of Genetics*, by Gardner et al., copyright ©1991 by John Wiley & Sons, Inc. Reprinted by permission by John Wiley & Sons, Inc.)

They are both double ring structures called **purines**. The structures of C and T are also similar, but with only a single ring. They are called  **pyrimidines**. When DNA is not at the replication stage, A and T and C and G are always bound together in order to keep the DNA molecule stable. These genetic codes are attached to a sugar-phosphate backbone, a polynucleotide chain which is composed of the sugar molecule 2-deoxyribose and phosphate.

(2)  The length of DNA molecules in a chromosome is in the range of a few milli-meters. When it is not in action (replication or transcription), it coils onto a protein called a **histone**. The chain coils many times around the histone so that the length of the string is reduced by $10^{-4}$ (see Fig 1.3.1). How the coiling is actually done is still a mystery.

(3)  Fig. 1.3.2 shows how a DNA molecule replicates itself during cell division. The replication is simply a series of chemical reactions when the conditions are right.

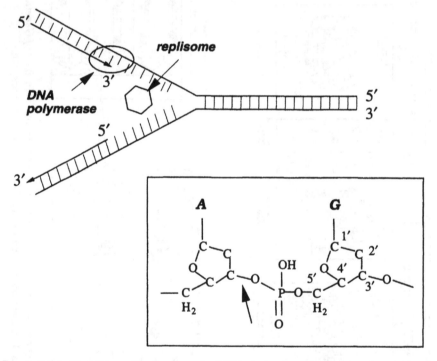

Figure 1.3.2: Top figure: Replication of a DNA molecule. The enzyme that unwinds the double stranded DNA is called replisome and the enzyme that replicate the single stranded DNA into double is called DNA polymerase. Lower right insert: The meaning of the 3′ and 5′ ends of a DNA molecule. The numbers are used to locate the carbon positions in the sugar base. Each genetic base stops or starts at the bond with the arrow sign. It is at the 3′ site of one base and the 5′ end of the another. A new unit can be added to the right (from 5′ to 3′), but not to the left, during DNA replication.

Several enzymes, including the DNA unwinding enzyme and building materials, are necessary for replication to take place. Note that the newly added strand always moves from the 5' end to the 3' end. The meaning of 5' and 3' is based on the positions of the carbon in the sugar base of the DNA.

(4) Three genetic codes, called **codons**, represent one amino acid. The DNA code for amino acid is given in Table 1.3.1. The U (uracil) replaces T in coding because amino acid is synthesized for mRNA where the T in DNA is transferred into U. Two leaders of this discovery, Nirenberg and Khorana, shared the Nobel Prize in 1968.

(5) How the DNA code is used to manufacture a protein is one of nature's wonders. The steps are shown in Fig. 1.3.3. The DNA is coverted to RNA (ribonucleic acid) for three different functions. The main component is the mRNA (m for messenger) which has the code for a protein. The tRNA (t for transfer) will form a molecule that on one side matches the DNA codes and on the other side matches the

Table 1.3.1: The genetic code. The English abbreviations, Phe, Leu, etc. are names for amino acids. Only **Stop** means a stop code that ends the protein synthesis.

|   | U | | C | A | | G | | |
|---|---|---|---|---|---|---|---|---|
|   | Phe | Leu | Ser | Tyr | **Stop** | Cys | **Stop** | Try |
| U | UUU | UUA | UCU | UAU | UAA | UGU | UGA | UGG |
|   | UUC | UUG | UCC | UAC | UAG | UGC | | |
|   | | | UCA | | | | | |
|   | | | UCG | | | | | |
|   | Leu | | Pro | His | Gln | Arg | | |
| C | CUC | | CCU | CAU | CAA | CGU | | |
|   | CUU | | CCC | CAC | CAG | CGC | | |
|   | CUA | | CCA | | | CGA | | |
|   | CUG | | CCG | | | CGG | | |
|   | Ile | Met | Thr | Asn | Lys | Ser | Arg | |
| A | AUU | AUG | ACU | AAU | AAA | AGU | AGA | |
|   | AUC | | ACC | AAC | AAG | AGC | AGG | |
|   | AUA | | ACA | | | | | |
|   | | | ACG | | | | | |
|   | Val | | Ala | Asp | Glu | Gly | | |
| G | GUU | | GCU | GAU | GAA | GGU | | |
|   | GUC | | GCC | GAC | GAG | GGC | | |
|   | GUA | | GCA | | | GGA | | |
|   | GUG | | GCG | | | GGG | | |

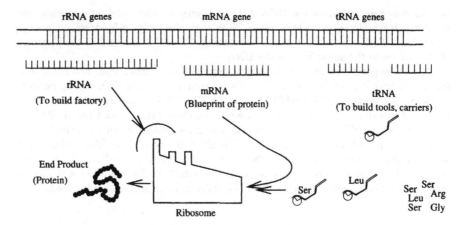

Figure 1.3.3: How protein is synthesized from genes. At least three major parts, rRNA, mRNA and tRNA, are needed. The mRNA is the blueprint of the specific protein to be synthesized, rRNA builds the ribosome, like a factory for a product, tRNA is a tool that reads the genetic code in mRNA and supplies the amino acid sequentially to the ribosome.

correct amino acid. The protein synthesis is done in a small organelle called a **ribosome**, which is made by the ribosome rRNA (RNA).

(6) Not all genetic codes on a chromosome are genes. A very large portion of DNA seems to serve no useful function. It appears that some portions of chromosome can be missing without affecting a creature's functions. They were once called junk DNA, but it has been recently found (in 1994) that they may serve a purpose, too. The nongene portion of the DNA is called **intron** and the gene parts are called **exon** (see Fig. 1.3.4). Note that the intron parts are deleted during RNA formation (see Fig. 1.3.5).

(7) The human haploid (single copy of chromosomes) contains approximately $3 \times 10^9$ codes or **base pairs** (bp, see Table 1.2.2). Gene sizes vary greatly depending on the proteins they make. They are usually in the $10^3 \sim 10^4$ bp range.

(8) If we put gene actions into a timely sequence, they act like a computer program. The way that a programmer tells a lifeless computer to do things is very similar to the way that nature tells the lifeless molecules to act. A simple expression would be:

|  |  |
|---|---|
| IF (SIGNAL **A** EXISTS) | THEN PRODUCE **B** |
|  | ELSE PRODUCE **C**; |
| IF (**B** AND SIGNAL **D** EXIST) | THEN PRODUCE **E** |
|  | ELSE DO NOTHING |
| REPEAT PRODUCING **C** UNTIL | (**A** EXISTS) |

In the above genetic program, **A** and **D** can be signals from other parts of the body, such as hormones or neural signals, or they can be outside signals, such as temperature,

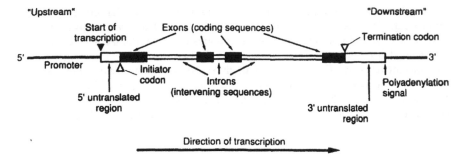

Figure 1.3.4: The general structure of a genome in the neighborhood of a gene. The transcription can start only when the promoter is activated by a proper enzyme. It ends with a long chain of $AAAAAA\ldots$, called polyadenylation signal. (From *Genetics in Medicine*, by Thompson et al. (1991), copyright ⓒby W. B. Saunders Company and the authors. Reprinted by permission.)

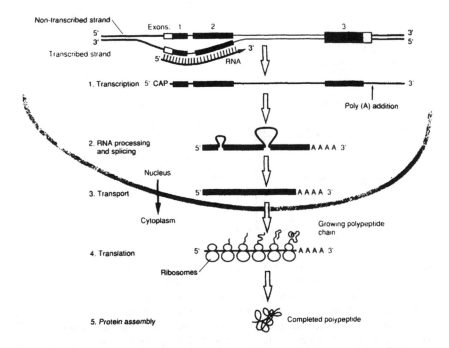

Figure 1.3.5: The flow of information from DNA to RNA to protein. Of the two strands of DNA, the one for RNA transcription is also called the anticoding (or antisense) strand and the nontranscription strand is also called the coding (or sense) strand. Special attention should be paid to the 3' and 5' directions of the two strands. (From *Genetics in Medicine*, by Thompson et al. (1991), copyright ⓒby W. B. Saunders Company and the authors. Reprinted by permission.)

pressure, bacteria or antigens; **B** and **C** must be something that the cell can produce from its genetic codes. The implementation of this type of program is given in Figs 1.3.6–7. This model was confirmed by one famous example, the lac gene of *E. coli*. *E. coli* can digest both glucose and lactose, but the two sugars have to be digested by two different enzymes. *E. coli* does not produce the two enzymes simultaneously. It produces the correct enzyme in response to the existing sugar. When the sugar changes, the enzyme changes. Because *E. coli* is relatively easy to work with, the intermediate chemicals, such as inducer and repression enzymes, could be identified

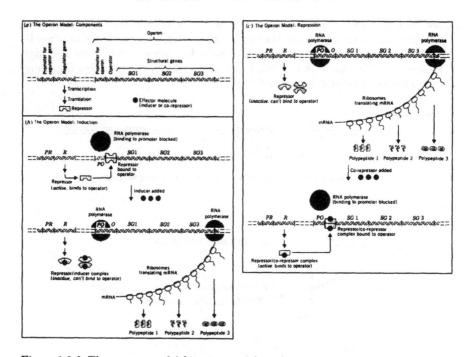

Figure 1.3.6: The operon model for structural (main) gene and regulator gene. (a) The structure of genes and how they are regulated by promoters. (b) The operation is: If (inducer exists) then produce polypetides. Repressors are sent out by the regulator gene. If they do not meet any inducer, they will block the structural gene by binding at its operator. The structural gene then is not functioning. On the other hand, if inducers are in the environment, they bind with the repressor. Now the RNA polymerase can bind with the structural gene's promoter and start the structural gene transcription. (c) The operation is: If (co-processor exists) then stop producing polypetides. This is the situation when an outside chemical co-repressor will suppress the protection of the protein. (From *Principles of Genetics*, by Gardner et al., copyright ©1991 by John Wiley & Sons, Inc. Reprinted by permission by John Wiley & Sons, Inc.)

(see Fig. 1.3.6). This experiment was done by F. Jacob and J. Monod in 1961 and they were awarded the Nobel Prize in 1965.

(9) Genes can be missing or mutate. In the most drastic case, all or part of the whole chromosome can be missing, or the number of chromosomes can also increase. We know that a person with XX chromosome is female and with XY is male, but there are individuals with X, XXY and XYY chromosomes. An abnormal numbers of chromosomes can be fatal or cause genetic diseases. (Persons with X or XXY are infertile, but persons with XYY are basically normal except some individuals may have behavioral problems.) Mutation on a single gene is very common. Most times, only one base pair in the genetic code is altered. This can cause great phenotypic change if the mutation is at a crucial site in the protein. For example, sickle cell anemia is caused by a single base pair mutation. But most times small mutations have

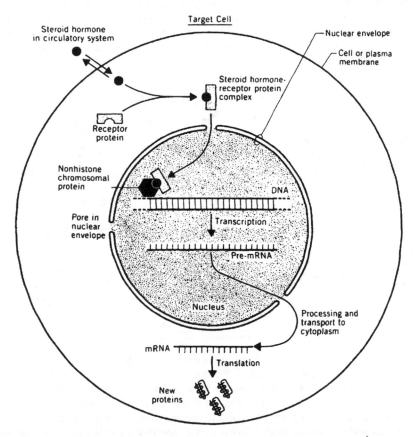

Figure 1.3.7: This diagram illustrates how extra-cellular signals regulate the cell function. (From *Principles of Genetics*, by Gardner et al., copyright ©1991 by John Wiley & Sons, Inc. Reprinted by permission by John Wiley & Sons, Inc.)

no noticeable effect. In fact, the proteins performing the same function can be different in their detailed amino acid compositions among individuals.

## 1.4   Tools in Molecular Genetics

### 1.4.1   Restriction Fragment Length Polymorphism (RFLP)

In §1.1 we have shown that a pedigree can be used to locate a gene relative to the location of another gene. But how do we get the position of the first gene? In animal experiments, such as those on fruit flies, we can study the effect caused by the deletion of a section of a chromosome. For example, when a section at the end of chromosome I was missing (see Fig. 1.2.5), we found bobbed-hairs abnormality in the phenotype. Thus, we know the position of this gene. Other genes can then be located by their

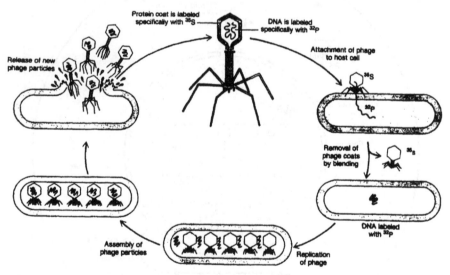

Figure 1.4.1: Life circle of virus. From top clockwise. (Top center:) (1, top figure) A virus. (2) It attaches to a cell. (3) its DNA is injected into the cell, but its protein coat is left outside the cell. (4, bottom figure) Viral DNA and the protein coats are produced in the cell. (5) Viral DNA and coats combined into complete viruses. (6) Cell bursts, releasing the virus particles. This picture was drawn to describe the experiment by Hershey and Chase who used different radioactive elements $^{35}$S and $^{32}$P in the coat and DNA and showed that the coat must be left outside the invaded cell. (From *Genetics*, by Weaver and Hedrick (1989), copyright © by The McGraw-Hill Companies. Reprinted by permission.)

linkage relations with this gene and their mutual linkage relations. However, this is a very time consuming process, because missing parts in chromosome do not occur often. We need a better tool to identify the locations of some "initial" genes. Actually, the contents in these locations do not have to be genes. They can be DNA fragments that are polymorphic and easily identifiable. In the Fig. 1.1.2 example, the serum proteins were used to locate the gene of *dentinogensis imperfecta*. Only two crucial properties of these serum proteins were useful in gene hunting: (1)they could be identified from every individual, and (2) the proteins had many alleles (polymorphism). Now if on the chromosomes there are **markers** that are easily identifiable and polymorphic, then the markers can serve the role of serum proteins in gene hunting. The method to find the exact locations of many polymorphic genetic markers is called restriction fragment length polymorphism (RFLP). It started with the relation between a virus and its prey bacteria.

The life cycle of a virus plays an important role in DNA manipulation. Fig. 1.4.1 shows the typical life cycle of a virus. In nature, a virus can invade a cell or a bacterium, but in the laboratory, bacteria are more commonly used as the prey. In the 1960s, researchers already noticed that certain bacteria were immune to the virus invasion. Apparently, the function of the injected virus DNA was "restricted" inside the bacterium. We now know that the bacteria can produce a chemical, called *restriction enzyme*, that can cut the virus DNA. This enzyme binds at a specific site on the DNA molecule and separates it. The function of a restriction enzyme, the endonuclease EcoR1, is given in Fig. 1.4.2. W. Arber (in the 1960s) and H. Smith (in 1970) were two of the early researchers of this process. They shared the Nobel Prize in 1978. More than 100 restriction enzymes have currently been discovered. Table 1.4.1 shows four

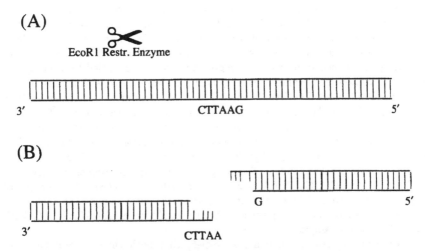

Figure 1.4.2: Function of restriction enzyme EcoR1. (A) When EcoR1 is added to a solution with a DNA molecule that happens to have a site with CTTAAG. (B) The DNA is cut at this site and any site with composition CTTAAG.

Table 1.4.1: Recognition sequences and cutting sites ($\updownarrow$) of four restriction endonucleases.

| Enzyme | Recognition sequence and cutting sites | | | | | | |
|--------|---|---|---|---|---|---|---|
| EcoR1 | G | | A | A | T | T | C |
| | C | $\updownarrow$ | T | T | A | A | G |
| HaeIII | | G | G | | C | C | |
| | | C | C | $\updownarrow$ | G | G | |
| PstI | C | T | G | C | A | | G |
| | G | A | C | G | T | $\updownarrow$ | C |
| SmaI | C | C | C | | G | G | G |
| | G | G | G | $\updownarrow$ | C | C | C |

commonly used restriction enzymes and their binding sites. The binding site for EcoR1 is CTTAAG. If we consider that the DNA codes are randomly arranged, each with probability 1/4 to appear, then the chance of meeting a site of this composition is roughly $1/(4^6) \approx 1/(4 \times 10^3)$. Thus, when a DNA strand is cut by this enzyme, the average length of the fragments is $4 \times 10^3$. If the human genome is cut by it, it will turn into roughly $3 \times 10^9/(4 \times 10^3) \approx 10^6$ fragments. Since the sites with composition CTTAAG are roughly randomly distributed in the whole genome, the cut produces fragments with various lengths (see Exercise 1.9). The fragments can be separated by electrophoresis according to their lengths and can be identified by a portion of their configuration. For example, one fragment may have a site with a piece of code sequence ATATGCAATTGGGGGAAATT. Under the random coding assumption discussed previously, the chance to match a 20 bp DNA sequence is only $4^{-20} \approx 10^{-12}$. Thus, this code sequence should determine a unique piece of a fragment in the human genome of length $3 \times 10^9$ bp. The detailed description of fragment identification is given in Fig. 1.4.3. In the hybridization procedure (8), the radioactive probes, which are short single-stranded DNA fragments of a given composition (such as TATACGTTAACCCCCTTTAA manufactured with radioactive $P^{32}$), are mixed with the single stranded long DNA fragments on the filter paper. The probes can only be attached to their complement DNA (it would be ATATGCAATTGGGG-GAAATT in this example).

Since chromosomes are in pairs, we expect that both chromosomes have one fragment that matches with this probe. As mentioned before, fragments cut by a 6 bp restriction enzyme have average length of $4 \times 10^3$ bp. Thus, a probe of length 20 bp occupies only a small portion on the restriction fragments. There is no guarantee that

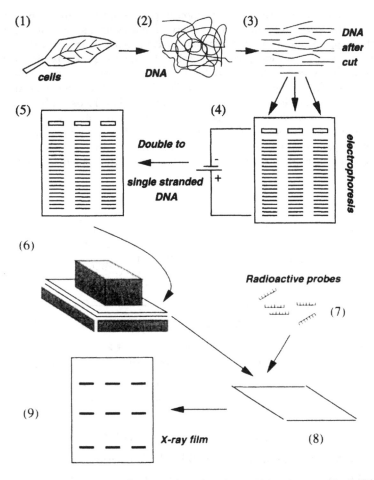

Figure 1.4.3: How restriction fragment length polymorphism is recognized. (1) Some living cells. (2) When washed by special chemicals, only the DNA in the cell nuclei is left. (3) When cut by a restriction enzyme, the DNA is now in fragments of different lengths (almost continuous in length). (4) When the fragments are put in an apparatus for electrophoresis, the longer fragments move slower than the shorter one in the gel. Hence they become separated. (5) Alkali is used to denature the double stranded DNA into single strands. (6) Southern Blot procedure: The gel is put between filter papers so that the single stranded DNA molecular are attached to a paper. (7) A special radioactive DNA probe (a single stranded DNA fragment with radioactive $P^{32}$ in its backbone). (8) These probes will attach only to the fragments that have their complements, called hybridization. (9) X-ray film shows only the fragments that matches with the probe. Usually the three (or more) slots of the gel in Step (4) are filled with DNA fragments from different specimens. Thus, the three slots of the film in (9) usually show different bands (see Fig. 1.4.4 and Fig. 2.1.3).

the rest of the portions of the two restriction fragments are the same.The fragments that match the same probe may have variety of lengths in the human population. When this happens, the restriction fragment length polymorphism (RFLP) can serve as alleles at a given locus for gene hunting. RFLP happens quite often in the intron part of the DNA where some small DNA segment, called tandem, may repeat itself many or hundreds of times such as GTGTGTGTGT.... The repetitions are called **variable number tandem repeats (VNTR**, also called **microsatellites**) with the notation $(GT)_n$ for $nGT$ repeats. These repeats cannot possibly serve as codon to produce useful protein, but it does not matter since they are located in the intron region. Why repeats happen is still unknown. Fig. 1.4.4 shows the usefulness of using VNTR as a genetic marker to reveal how a particular VTNR is inherited in a family. It

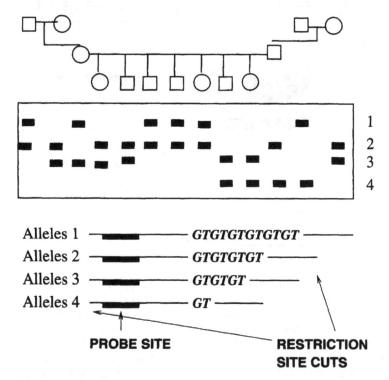

Figure 1.4.4: Using RFLP marker to trace the inheritance route of gametes in family members. The top figure is the family pedigree, followed by their gel bins associated with a particular restriction endonuclease enzyme and probe. The bottom figure explains the reason for the bin polymorphism. The probe picked the "same" piece (locus) of DNA fragment in the genome of the family members. However, due to the GT repeats, their lengths vary. If a gene is in the neighborhood of this marker, we can predict the gene's paths in the family.

takes effort to identify a probe for RFLP. Thus, many probes have to be purchased, but many others are in the public domain.

## 1.4.2  Polymerase Chain Reaction (PCR)

Polymerase Chain Reaction (**PCR**), discovered by Saiki et al. (1986), completely revolutionize DNA amplification. Previously, DNA molecules could only be amplified by first transporting a DNA molecule into a bacterium and then letting the DNA multiply as the bacteria multiplied. This is a labor-intensive process as we will see in the next section. PCR now enables a single DNA or RNA to amplify several million-fold in a few hours. The key to this method is to use two primers to let the DNA molecules replicate *in vitro* as if it were *in vivo*. A primer is a piece of single stranded DNA that can initiate replication when it attaches itself to another piece of single stranded DNA in the right environment. The details of this procedure are given in Fig. 1.4.5. The mathematical side of this process is given in Fig. 1.4.6. Only the piece of DNA that matches the two primers at the two ends will be amplified. Suppose each primer contains 20 basepairs. Then the chance that a piece of DNA can match with

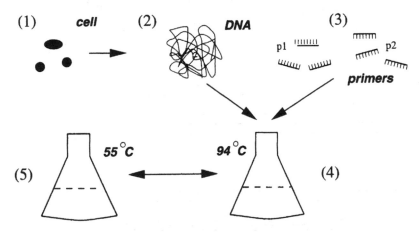

Figure 1.4.5: The PCR procedure. (1) Any cells, such as drops of blood. (2) When the proteins are washed away, only the DNA is left. (3) Two primers, p1 and p2, and the DNA are mixed in a DNA replication environment. ( p1 can be the same as p2.) (4)–(5) The two environments for DNA molecular chain reaction. At stage (4) the solution is heated to 94°C. The double-stranded DNA becomes single-stranded. At stage (5) the solution is cooled down to 55°C so that the single-stranded DNA is replicated by the two primers. The timing is important in this iteration process. You do not want the time to be so short that the replication has not reached the other end, or so long that material is wasted in the unwanted portion. Sometimes three temperature ranges are used. DNA replication is set at 72°C. See the mathematical side of this process in Fig. 1.4.6.

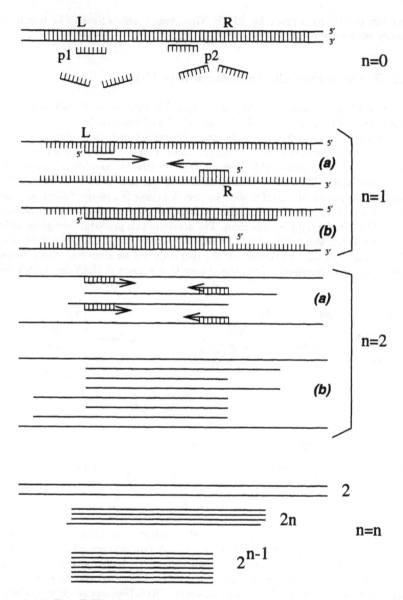

Figure 1.4.6: How PCR works. At time $n = 0$, DNA molecules and primers are mixed in the right environment for DNA replication. Primer $p1$ matches the $L$ side of the DNA and $p2$ the $R$ side. At $n = 1$, the two primers begin to replicate ((a) step). The process stops by heating to denature the double-stranded DNA ((b) step), When the process is repeated $n$ times, we will have $2^{n-1}$ fragments of the DNA between $L$ and $R$ ends of the original DNA.

both is $4^{-40} \approx 10^{-24}$. For most genomes it is a unique piece. The primers can be custom made. The current price is a few dollars per base pair.

Note that this method can start with a very small amount of DNA, less than 1% is required by RLFP, which in order to produce any noticeable band in gel has to have a large amount (in molecular scale). If the DNA section between the two end primers

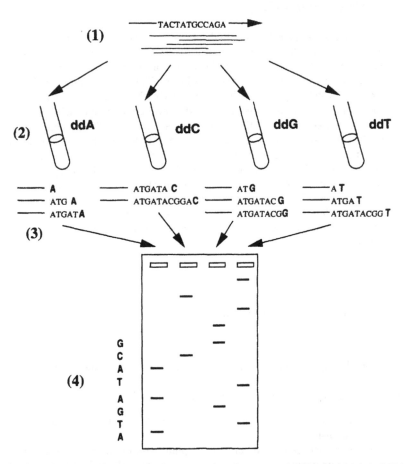

Figure 1.4.7: The Sanger method of DNA sequencing. (1) Single-stranded DNA molecule. The example is a primer attached to TACTATGCCAGA .... (2) They are put into four test tubes with DNA replication chemicals and each with one stopping enzyme. (3) Due to randomness in meeting the stopping enzyme during the replication process, the products in each tube have fragments with variable lengths. However, all fragments stop at the same DNA base in the same tube. (4) When the production undergoes electrophoresis, the fragments are separated by weights. Hence the original DNA sequence can be determined. For example, the lightest fragment must end in A, the second lightest end in T, etc. (Redrawn from Weaver and Hedrick, 1989.)

has polymorphism in length, then this section forms a good marker. This method is particularly desirable in forensic science because the evidence in a crime can be as minute as a drop of blood or a single strand of hair. We will see this application later.

### 1.4.3  DNA Sequencing

The purpose of DNA sequencing is to find the order of the genetic codes A, T, C, G in a given fragment of DNA. The key chemical for this method is called **dideoxyribonucleotide** (note DNA is deoxynucleotide) which can stop DNA synthesis. There are four types of dideoxynucleotides, labeled ddATP, ddTTP, ddCTP and ddGTP, which can respectively stop the synthesis when they are attached to A, T, C and G. In DNA sequencing four test tubes are needed. All test tubes are supplied with DNA replication chemicals, but each with a different stopping agent dd—. The procedure is shown in Fig. 1.4.7. Not all sequences stop at the same point because of the randomness in the time when the dd— found a matching site. When the single-stranded parts are washed away, what is left are fragments with different lengths. The gel picture of Fig. 1.4.7 shows the relation between stopping codes and the lengths of the fragments. This method was discovered independently by W. Gilbert and F. Sanger who shared the 1980 Nobel Prize for their discovery. Since a gene usually contains several thousand base pairs, this is still a time-consuming process. Scientists are still trying to find quicker methods for DNA sequencing.

## 1.5    Genetic Engineering

The main goal of genetic engineering is to create new plants and animals that have new characteristics.

Two key technologies in genetic engineering are:

1. To fabricate a gene with a given code sequence or to cut a desirable gene from a living organism.
2. To transport the gene to an animal's or plant's reproductive cell and make sure the new animal or plant can express the gene correctly and pass the trait on to the future generations.

To fabricate a piece of DNA with a given sequence of codes can be done. Its shorter versions, the primers or probes, are custom-made regularly. However, for a long sequence of amino acids, we still do not have enough knowledge to predict how they will fold into a three-dimensional protein and what function this protein will have (see Richards, 1991). Thus, at present, most desirable gene are cut from the genome of a living creature. Cutting out a gene from a genome is relatively easy if the gene's location is known. We do not even need to sequence the gene. If we know that a gene is between two probes $p1$ and $p2$ (see Figs 1.4.5–6), then the gene can be automatically "cut" by the two primers and magnified by PCR (see Fig. 1.4.6). Between the two probes there may be more DNA than the gene itself. This may not matter because the other portion may be junk, or some gene that may not function when transplanted. Of course, the desirable gene may not work after transplantation either. A lot of trial and error is involved at this state of the art. Fig. 1.5.1 shows how a piece of mouse DNA

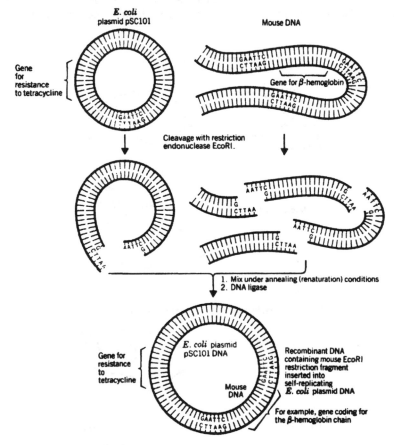

Figure 1.5.1: How a piece of mouse DNA is inserted into a bacterial plasmid *in vitro*. Note that the restriction endonuclease EcoR1 cut the *E. coli* and mouse DNA at the same site. When they are mixed, the mouse DNA may inserted into the *E. coli* plasmid. The *E. coli* GAATTC site should be part of another antibiotic-resistant gene such as ampicillin resistant. With these two antibiotic-resistant sites, we can easily recognize which bacteria have the mouse DNA inserted into them. See Fig. 1.5.3 for details. (From *Principles of Genetics*, by Gardner et al., copyright ©1991 by John Wiley & Sons, Inc. Reprinted by permission by John Wiley & Sons, Inc.)

can be transported into *E. coli* **plasmid** ps101. A plasmid is a closed DNA loop that
**may** act like a chromosome in the host cell. A wide range of different plasmids exist in
the nuclei of bacteria. They are part of their genetic material. If the environment is
correct in a host, a plasmid will reproduce itself and then produce the protein it
represents. Note that the real plasmid is not a circle like the one drawn in the figure.
There is great flexibility on the length of the new DNA to be implanted.

Fig. 1.5.2 shows four commonly used methods to introduce DNA fragments into a
cell. The first method is to let the plasmid fuse into the cell using specific chemical
treatment. The cells are then cultured in another medium so that the membrane will

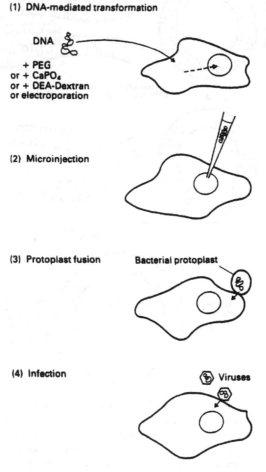

Figure 1.5.2: Four methods of introducing DNA to cells. (From *Genetic Engineering*,
by Kingsman and Kingsman (1988), copyright ©by Blackwell Scientific Publications
Ltd. Reprinted by permission.)

recover from its wound. Once the DNA is taken up into the cell cytoplasm, there is only a small chance that it can make its way to the nucleus. This method is called **DNA-mediated transfection**. Fig. 1.5.3 shows how to confirm that foreign DNA has been inserted into a particular bacterium. This method was used to replicate DNA fragments before PCR was invented. Greater effort was needed to do this experiment than PCR. However, using bacteria or yeast to replicate long DNA is still a common procedure. Usually PCR can only replicate short DNA fragments. For very long DNA fragments, PCR tends to end with errors due to the imperfection of repeated denaturing and replicating.

The second method in Fig. 1.5.2, called **micro-injection**, is a high-tech method of directly injecting DNA into the cell nucleus (see Fig. 1.5.4). This method has been used to produce transgenic animals. An early experiment by Gordon et al. (1980) used

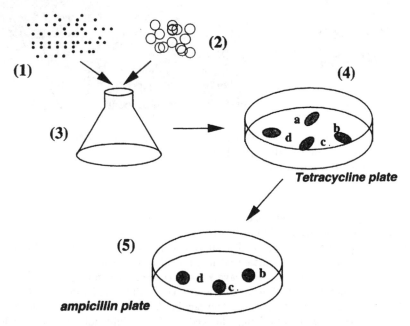

Figure 1.5.3: How to recognize a mouse DNA inserted bacteria (*E. coli*). (1) Wild-type *E. coli*, no resistance to either antibiotic, tetracycline or ampicillin. (2) Plasmid from Fig. 1.5.1. The original plasmid can resist tetracycline and ampicillin, but the ampicillin-resistant gene is destroyed if the mouse gene is inserted. (3) When they are mixed with certain chemicals, some of the plasmids may permeate through the cell membranes into the bacteria. (4) When the bacteria are cultured on a tetracycline plate, only those with the inserted plamid can survived. (5) And when they are put onto an ampicillin plate, the colonies that do not survive must be the colonies with the mouse gene, namely colony *a* in which the ampicillin gene must have been replaced by the mouse gene.

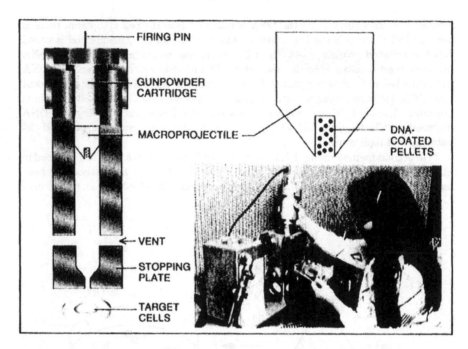

FIRING PIN

GUNPOWDER
CARTRIDGE

MACROPROJECTILE

DNA-
COATED
PELLETS

VENT

STOPPING
PLATE

TARGET
CELLS

Figure 1.5.4: The structure of a DNA-Particle gun used to inject DNA molecules into cells. (From Gasser and Fraley (1992), copyright ©by John C. Sanford and Laurie Grace. Reprinted by permission.)

mice (see Fig. 1.5.5). Fertilized eggs of a mouse were used. Foreign DNA was injected into the male **pronucleus** (haploids from sperm) because the male pronucleus is larger than the female pronucleus. The surviving eggs were then introduced into the uterus of a pseudo-pregnant female that had mated with a vasectomized male. Many transgenic mice with human genes have since been produced, such as the ones with human growth hormone or fat gene (Thompson, et al., 1991 and Pennisi, 1994, 1997).

There are many ways to produce transgenic plants. It is easier to regenerate a plant from its cells than it is to change the seeds. With the right culture medium, a leaf cell can grow into a whole plant (Fig. 1.5.6). Culture media contain nutrients and growth hormones, but their compositions and concentrations are the key elements of success. They vary from plant to plant and not every plant can be regenerated by this method. Another method, called **protoplast culture**, is done by first removing the cell wall of the plant cell with a mixture of cellulose and pectinase enzyme. A plant cell with its cell wall removed (called a protoplast cell) can be treated as an animal cell. DNA can be introduced by transfection or microinjection. Again with the right medium, the protoplast can regenerate its wall and grow into a new plant.

DNA fragments can be carried into cells by bacteria or viruses. These vehicles are called **vectors**. Two commonly used gene transfer vectors in plant gene transports are

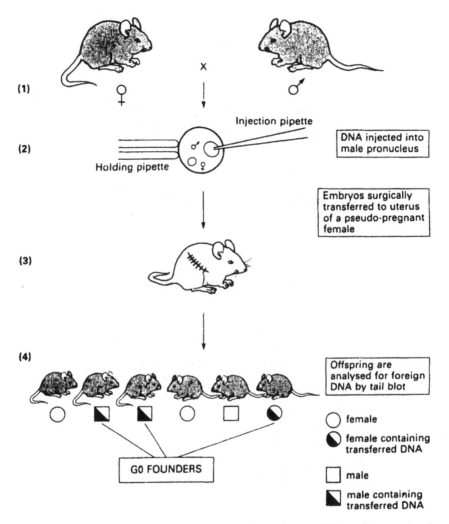

Figure 1.5.5: How a transgenic mouse is produced. (From *Genetic Engineering*, by Kingsman and Kingsman (1988), copyright ©by Blackwell Scientific Publications Ltd. Reprinted by permission.)

the bacteria *Agrobaterium tumefacien* and *Agrobaterium rhizogenes*. They can infect over 300 different plants through wounded tissues. Some of these bacteria carry plasmids. When an *A. tumefacien* carries a *Ti* plasmid, it causes the wounded tissue to proliferate into a crown tumor (see Fig. 1.5.7). The *Ti* plasmid, like other plasmids, can be used to transport foreign DNA. It acts in the way we have seen in Fig. 1.5.1. When the foreign DNA replaces the tumor DNA in *Ti*, the plasmid can no longer

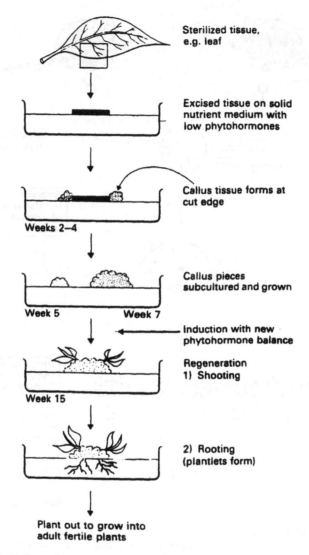

Figure 1.5.6: Callus culture to generate a plant from a leaf cell. (From *Genetic Engineering*, by Kingsman and Kingsman (1988), copyright ©by Blackwell Scientific Publications Ltd. Reprinted by permission.)

produce the tumor. The foreign DNA may be expressed in the new plant. One successful transgenic plant was created by Vaeck et al. (1987) where the bt-gene that produced a toxic chemical from a bacterium *B. thuringiensis* was transferred to a tobacco plant. Caterpillars that thrive on normal tobacco plants could no longer

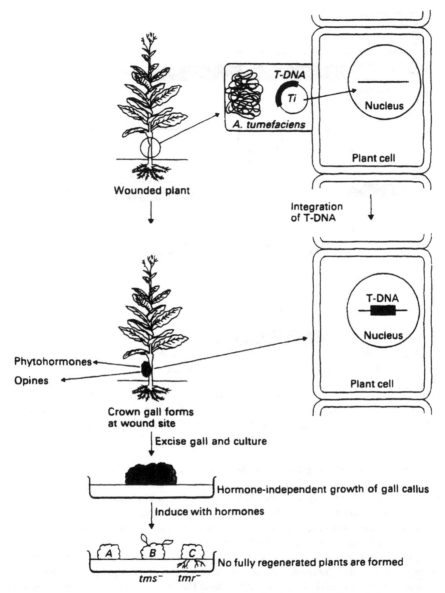

Figure 1.5.7: The interaction of *A. tumefaciens* with plants. (Top figure) A wound site is infected with *A. tumefaciens* which carried a special T-DNA in a plasmid. (Middle figure) When the plant is infected, the T-DNA is transferred to the plant cell nucleus in the wound repair cell. (Bottom figure) The plant cell with T-DNA can form a new plant in the right culture. (From Kingsman and Kingsman (1988), copyright ©by Blackwell Scientific Publications Ltd. Reprinted by permission.)

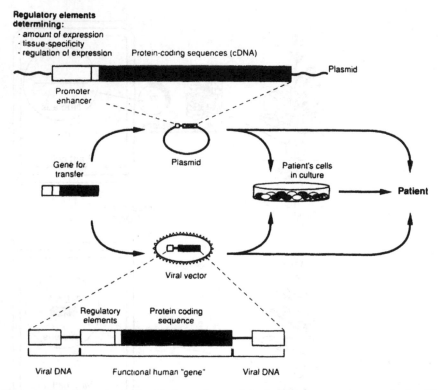

Figure 1.5.8: Two strategies that can be used to transfer a gene to a patient with genetic disease. (From *Genetics in Medicine*, by Thompson et al. (1991), copyright ©by W. B. Saunders Company and the authors. Reprinted by permission.)

survive on the transgenic plant due to the toxin. Another very impressive experiment was the successful transfer of a firefly luciferase gene to the tobacco plant. The resulting plant glowed with green light in the dark (see Weaver and Hedrick, 1989, p. 467 for a beautiful color picture). This shows that the basic genetic functions of plants and animals are still transferable despite more than 500 million years of separation.

DNA fragments can also be transferred to cells by viruses. By studying the life cycle of a virus (see Fig. 1.4.1), it is intuitive that a virus can be used as a gene transporter if a DNA fragment can be inserted into the virus. The idea is to insert one section of the virus DNA with a specific gene is the same as the recombinant method described in Fig. 1.5.1−2. Once this piece of recombinant DNA is inserted into a cell and once this cell is invaded by the virus, the progeny may contain viruses with the desired recombinant DNA. The viruses can then be used to invade other cells one wishes to have this DNA transmitted into.

Fig. 1.5.8 shows how gene transfer can be used to treat genetic disease or cancer. In this procedure the culture of human cells *in vitro* is crucial, but we have no problem in doing it now. There are currently more than 100 clinical trials using gene therapy (see Marshall, 1995).

# Exercise 1

**Note: These exercises assume that the reader has a background of one year of graduate work in statistical theory. The statistical methods to solve these problems may not have been mentioned in the chapter.**

1.1 The following pedigree (Fig. E1.1)came from a Venezuelan family with Huntington's disease (a dominant gene with manifestations of muscle movement problem, speech difficulties and brain tissue degeneration). The marker to locate this gene has three alleles $A$, $B$, and $C$. Use the family with eight children at generation V only to construct the likelihood function for the recombination fraction $\theta$ between the disease gene and the marker. Is it more likely for $\theta = 0$ or $1/2$?

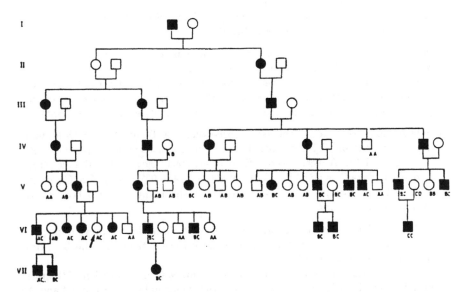

Figure E1.1. Source: Gusella et al. (1983), copyright by Macmillan Magazines, Ltd and the authors. Reprinted by permission.

1.2 A general selection model for two alleles $\{A, a\}$ can be formulated as follows. Let the relative survival rates for $AA$, $Aa$ and $aa$ be $\sigma_1$, $\sigma_2$ and $\sigma_3$ ($\sigma_1 + \sigma_2 + \sigma_3 = 1$).

Suppose the initial frequencies of $AA$, $Aa$ and $aa$ are $p_0^2$, $2p_0q_0$ and $q_0^2$ $(p_0 + q_0 = 1)$.
(i) Show that in the next generation by random mating, the proportions of AA, Aa and aa can still be written as $p_1^2$, $2p_1q_1$ and $q_1^2$ $(p_1 + q_1 = 1)$, with

$$p_1 = (\sigma_1 p_0^2 + \sigma_2 p_0 q_0)/\omega,$$
$$\omega = \sigma_1 p_0^2 + 2\sigma_2 p_0 q_0 + \sigma_3 q_0^2.$$

(ii) Let $p_n$ denote the proportion for the frequency of allele $A$. What happens to $p_n$ when $n \to \infty$?
(iii) In the past, there were governmental programs to eradicate bad recessive genes. Let this gene be denoted by $a$. One method is to prevent $aa$ persons from having children ($aA$ phenotype is indistinguishable from $AA$). Suppose the bad gene $a$ has a frequency of 0.01 in the general population and suppose one generation is 30 years. How many years would it take to reduce the $a$ allele frequency to half its current prevalence by this method?

1.3 Let a disease causing autosomal recessive allele have a frequency of 0.001 in the general population. A person with this disease marries a healthy person. What is the chance their next child will have the disease if:
(i) their first two children are one healthy and one with the disease;
(ii) their first two children are both healthy?

1.4 A famous experiment by Bateson and Punnett produced the following results. When two white-flowered sweet peas were crossed, the $F_1$ generation had all purple flowers. When the $F_1$ were intercrossed by self pollination, the $F_2$ generation had the ratio of purple and white flowers was $9:7$. Suppose the original peas were purebreds. Use the albino example to formulate a similar explanation.

1.5 Human blood types are determined by three alleles, A, B and O, in which A and B are codominant and O is recessive. Thus we have four blood types (phenotypes), A, B, O, and AB. A recent study (K. Roeder, 1994) showed that the bloodtype frequencies were $O : A : B : AB = 0.431 : 0.422 : 0.114 : 0.037$ among Yale Caucasian male students (sample size $n = 1000$). Have the frequencies reached Hardy-Weinberg equilibrium?

1.6 (i) It is well known among statisticians that most of Mendel's data were too good to be true, i.e., the ratios in Table 1.2.1 were too close to the ideal value $3:1$. How do you confirm this suspicion?
(ii) What is really suspicious about Mendel's data was his estimation of the ratio of $AA$ versus $Aa$ in $F_2$. According to his theory it should be $1:2$. Since $AA$ and $Aa$ had the same phenotype, Mendel had to do further experiment to ascertain them. He made his decision based on the phenotypes of their self-fertilized offspring ($F_3$). Plants of $AA$ would produce only phenotype $\mathbf{A}$, but plants of $Aa$ would produce types $\mathbf{A}$ and $\mathbf{a}$. This is easy to determine for traits related to the seed because one plant could produce a large number of seeds, but for the traits related to the flower and pod, he had to wait until the seed grew into a full-grown plant. This took time

and intensive labor. He decided to plant 10 seeds from each of the 100 $F_2$ **A** plants (that meant 1000 plants) for the $F_3$ observations. His data were as follows:

| Trait | AA | Aa | Total |
|---|---|---|---|
| Petal color | 36 | 64 | 100 |
| Pod shape | 29 | 71 | 100 |
| Unripe pod color(2 exp.) | 75 | 125 | 200 |
| Flower position | 33 | 67 | 100 |
| Stem length | 28 | 72 | 100 |
| Total | 201 | 399 | 600 |

The ratio is indeed very close to $1:2$. However, R. A. Fisher pointed out that the theoretical ratio should have been $1:1.696$ when Mendel's law was true. How did Fisher arrive at this ratio?

1.7 Let A, B and C be three consecutive adjacent loci on a chromosome. The distance between A and B is 0.2 Morgan and between B and C is 0.3 Morgan. Suppose the two distances 0.2 and 0.3 represent the crossover probabilities between A and B and B and C, and the two crossovers happen independently.
(i) Show that the crossover probability between A and C is not 0.5.
(ii) Show that Haldane's map function gives the correct probability for the A-C crossover.

1.8 In the previous exercise, we have

$$\theta_{AC} = \theta_{AB} + \theta_{BC} - 2\theta_{AB}\theta_{BC}, \qquad (E1.8)$$

where $\theta_{AC}$ is defined as the recombination fraction between loci $A$ and $C$, and $\theta_{AB}$ and $\theta_{BC}$ are defined in the same manner. However, it has been observed that this rule does not always hold. For example, there are cases where if there is a crossover between loci A and B, no crossover between B and C is possible. In this case Eq. (E1.8) fails. This phenomenon is called **interference**. The coefficient of interference is defined as

$$I = \frac{\theta_{AC} - (\theta_{AB} + \theta_{BC} - 2\theta_{AB}\theta_{BC})}{2\theta_{AB}\theta_{BC}}.$$

Show that the probability that there is a recombination in A and B but not in B and C is $\theta_{AB}[1 - (1 - I)\theta_{BC}]$.

1.9 If EcoR1 is used to cut the human genome, approximately how many of the fragments are longer than 10,000 bp?

1.10 The dimensions in Fig. 1.3.1 are quite accurate. Use those dimensions to find the approximate length of a chromosome. Consequently, how many folds a chromosome needs to magnified so it can be seen in the 1 cm range under a microscrope? (Note: $1 \text{ Å} = 10^{-8}$ cm)

1.11 Waaler in 1927 studied a large number of children in Norway for colorblindness. The following data were observed.

| Gender | Colorblind | Normal | Total* |
|--------|------------|--------|--------|
| Male   | 725        | 8324   | 9049   |
| Female | 40         | 9032   | 9072   |

Do the data agree with a sex-linked trait? If not, what other possible reasons can you suggest? (*From Spieces, 1977.)

# Chapter 2

# Linkage Analysis with Qualitative Trait

## 2.1 Basic Concepts in Pedigree Analysis

There are two basic approaches to making inferences on the recombination fraction $\theta$ between loci from a pedigree; the Bayesian and the likelihood approach. In both approaches, the likelihood function plays a key role. Let $F$ represent the pedigree information and $\theta$ (or $r$) be the recombination fraction between two loci. The likelihood function $L(\theta|F)$ is defined as the probability of having this (these) pedigree(s) when the recombination fraction is $\theta$. In (1.1.1), we have already seen that for a given $\theta$, the probability of having generation III in Fig. 1.1.3 can be obtained using a binomial distribution.

The notation in Fig. 2.1.1 is used to denote the relation between the alleles $\{A, a\}$ in the upper locus and alleles $\{1, 2\}$ in the lower locus. Without further information, the relation between the alleles can be one of the two following **phases**.

$$
\begin{array}{ccc}
\begin{array}{c}
| \quad | \\
\text{A}-| \quad |-\text{a} \\
| \quad | \\
2-| \quad |-1 \\
| \quad |
\end{array}
& \text{or} &
\begin{array}{c}
| \quad | \\
\text{A}-| \quad |-\text{a} \\
| \quad | \\
1-| \quad |-2 \\
| \quad |
\end{array}
\end{array}
$$

Figure 2.1.1: Notation for relation between markers in two loci.

The phase information in Fig. 2.1.1 can also be written as A2/a1 or A1/a2. Note that the phase information cannot be determined by laboratory tests on the individual alone, but it may be determined using information about the individual's ancestors. Fig. 2.1.2 shows that the genotype of the daughter must be A1/a2.

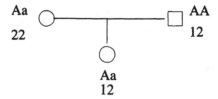

Figure 2.1.2: An example of phase determination by inheritance lines.

If an individual has only the information $(A, a)$, $(1, 2)$ without any other information from the ancestors to determine the phase, we may assume the two phases $A1/a2$ and $A2/a1$ have equal prior probability $1/2$.

## 2.1.1 Bayesian Inference on $\theta$

A Bayesian approach would not produce controversy if the prior can be reasonably assumed. In pedigree analysis, the prior can be reasonably assumed. Here we will use human pedigree as an example. Suppose the trait is not sex-linked. Since humans have 22 autosomes, it is then reasonable to assume that the recombination fraction $\theta$ between any two loci has a prior distribution

$$h(\theta) = \begin{cases} 1/2 & \text{with probability } 21/22 \\ \text{uniform in } [0, 0.5] & \text{with probability } 1/22. \end{cases} \tag{2.1.1}$$

When the two loci are on the same chromosome, we have the second prior, otherwise it is the first prior $1/2$. Since the location of the marker is usually known, a minor modification of (2.1.1) based on Table 1.2.2 can make the prior distribution more precise. For example, if the marker in on a shorter chromosome, then $h(\theta) = 1/2$ is larger than $21/22$. We leave the details of this modification as an exercise.

Let a pedigree $F$ be given. Then the posterior distribution of $\theta$ is

$$P(\theta|F) = \frac{L(F|\theta)h(\theta)}{\int_{0 < \theta < 0.5} L(F|\theta)u(\theta)d\theta + L(F|\theta)h(\theta = 1/2)}, \tag{2.1.2}$$

where $L(F|\theta)$ is the likelihood value for the pedigree given $\theta$, and $u(\theta)$ is a uniform distribution with density

$$u(\theta) = \begin{cases} 1/11 & \text{for } 0 < \theta < 1/2 \\ 0 & \text{otherwise,} \end{cases}$$

so that the total integral is $1/22$. To simplify the notation, let

$$L(\theta) = L(F|\theta), \text{ and}$$
$$L^*(\theta) = L(\theta)/L(\theta = 1/2).$$

Note that $L^*(\theta)$ is the likelihood ratio of some $\theta < 1/2$ versus $\theta = 1/2$. Substituting (2.1.1) into (2.1.2), the latter becomes

$$P(\theta|F) = \frac{22L^*(\theta)h(\theta)}{2\int_0^{0.5} L^*(t)dt + 21}.$$

If we define $\Lambda = 2\int_0^{0.5} L^*(t)dt$, then

$$P(\theta < 1/2|F) = \Lambda/(\Lambda + 21). \qquad (2.1.3)$$

The value of (2.1.3) shows the posterior probability that the two loci are linked. We can also calculate any posterior probability $P(\theta < \theta_0|F)$ for any given $\theta_0$.

**Example 2.1.1** *In Fig. 1.1.3, find the posterior probability of $\theta$ without generation I.*

**Sol**: By (1.1.2), $L^*(\theta) = [\theta^3(1-\theta)^7 + \theta^7(1-\theta)^3]/0.5^9$.

$$\Lambda = 2^{10}\int_0^{0.5}[\theta^3(1-\theta)^7 + \theta^7(1-\theta)^3]d\theta$$

$$= 2^{10}\int_0^1 \theta^3(1-\theta)^7 d\theta$$

$$= 0.776.$$

The derivation of the second equality is by a change of variable in the second term with $t = 1 - \theta$. Thus,

$$P(\theta < 1/2) = \Lambda/(\Lambda + 21) = \frac{0.776}{0.776 + 21} = 0.036.$$

This value is too small to claim $\theta < 1/2$. $\qquad\qquad\square$

**Example 2.1.2** *Suppose there is 1 recombination out of 12 meioses. What is the Bayesian estimate of $P(\theta < 1/2)$?*

**Sol**: $L(\theta) = 12\theta(1-\theta)^{11}$. $\Lambda = 2^{13}\int_0^{0.5}\theta(1-\theta)^{11}d\theta = 2^{13} \times 0.0064 = 52.043$. We have $P(\theta < 1/2) = 52.043/(52.043 + 21) = 0.714$. $\qquad\qquad\square$

**Example 2.1.3** *Suppose $n$ meioses were observed without recombinations, what $n$ do we need so that we have $P(\theta < 1/2) \approx 0.95$?*

**Sol**: $L(\theta) = (1-\theta)^n$, $\Lambda = 2^{n+1}\int_0^{0.5}(1-\theta)^n d\theta = 2^{n+1}(1 - 2^{-(n+1)})/(n+1) \approx 2^{n+1}/(n+1)$. Solve

$$\frac{2^{n+1}/(n+1)}{2^{n+1}/(n+1) + 21} = 0.95,$$

numerically, we have $\Lambda = 0.942$ for $n = 11$ and $\Lambda = 0.968$ for $n = 12$. In other words, even if two loci are very close, we need a sample of size at least 11 meioses to confirm linkage. $\qquad\qquad\square$

## 2.1.2   Likelihood Ratio Inference and LOD score

Based on the likelihood ratio test principle (see §A.2), we should use

$$Q \equiv -2\ln\frac{L(\theta = 0.5)}{\max_\theta L(\theta)} \sim \chi_1^2 \qquad (2.1.4)$$

to test the null hypothesis $H_0$: $\theta = 1/2$. Since the alternative hypothesis is one-sided $H_1$: $\theta < 1/2$, the asymptotic distribution of $Q$ is a $0.5:0.5$ mixture distribution of $\chi_1^2$ and a degenerate point $\{0\}$. (see (A.2.6) in Appendix A.2). Since it is more convenient to use a positive magnitude with base 10 logarithm, the LOD (logarithm of odds) score defined by

$$LOD = \log_{10}\frac{\max_\theta L(\theta)}{L(\theta = 0.5)}$$

is usually used in genetics. Note that other notations such as $LOD_{max}$, Lod, or lod are also used. By changing the log bases, it can be easily seen that (2.1.4) is equivalent to

$$4.6 \times LOD \sim \frac{1}{2}\chi_1^2 + \frac{1}{2}\{0\}. \qquad (2.1.5)$$

There is another notation Z for LOD at any given value $\theta$, i.e.,

$$Z = Z(\theta) = \log_{10}\frac{L(\theta)}{L(\theta = 0.5)}, \quad Z_{max} = \max_\theta Z(\theta) = Z(\hat{\theta}) = LOD,$$

where $\hat{\theta}$ is the $\theta$ that maximizes $Z(\theta)$ or $L(\theta)$. Usually, $LOD \geq 3$ is the minimum requirement to claim linkage. Note that the chance for $LOD \geq 3$ under $H_0$ is

$$0.5Pr\{\chi_1^2 \geq 4.6 \times 3\} = 0.5Pr\{\chi_1^2 \geq 13.8\} \approx 0.0001.$$

Why such a lower significance level ($10^{-4}$)? Because when the location of the gene to be searched is unknown, we usually have to try to link a disease gene with many markers. Hence this is a multiple decision problem and the false-alarm rate 0.05 for one search will not guarantee a 0.05 false alarm rate for many searches. Since the markers may also be linked, to work out the exact multiple decision false-alarm rate is not easy. The following is a well accepted argument for using $LOD \geq 3$ as the threshold in the initial stage of human gene localization.

By Bonferroni inequality, if $m$ markers are checked simultaneously, the overall probability of making a wrong claim of linkage in $m$ tests cannot be more than $m\alpha$ if the false-alarm rate in one test is no more than $\alpha$. When 20 markers are used in each chromosome, it is considered dense enough for the first stage linkage analysis. A denser marker list will usually give redundant results because a new marker will be very close to some of the existing markers. If we consider the number of markers is $20 \times 22 = 440$, then the LOD $\geq 3$ threshold can guarantee the wrong claim rate to be smaller than $440 \times 10^{-4} = 0.044$, a level close to the usual 0.05 significance level. Also, if we check this value with the Bayesian approach, which already considers the number of chromosomes in the prior, we see from Example 2.1.3 that we require $\log_{10} L^*(\theta = 0) = \log_{10} 2^{11} = 3.3$, not very far from the threshold LOD $\geq 3$. Lander and Botstein (1989) did some simulation study on this problem and concluded that when LOD $\geq 3$ is used the overall significance level is less than 0.05 for any number of markers (see §3.2 below).

A very desirable property of the $Z(\theta)$ score is that it is additive with independent families. Suppose there are $n$ families with pedigree information $F_1, F_2, \ldots, F_n$, then the joint Z function is

$$Z(\theta) = \sum_{i=1}^{n} Z_i(\theta),$$

where $Z_i(\theta)$ is the Z-function for family $i$. The LOD score obtained by the binomial distribution is relatively easy to compute for a large number of small families, but it is not suitable for a complex pedigree such as the one in Fig. 1.1.2 or Exercise 1.1. We will discuss them in the next section. The following is an example of small family study. We can see that the LOD score can be easily computed by inspection with the help of a simple computer program.

**Example 2.1.4**   *Bodmer et al. (1987) tried to locate the gene for special case of colorectal cancer,* familial adenomatous polyposis (FAP), *which affects an individual by developing hundreds or thousands of adenomatous polyps in the large bowel. Evidence shows that the disease is apparently inherited through a dominant gene. Fig. 2.1.3 shows the pedigree of one infected family. The marker used is C11p11 at chromosome 5 region 5q21-22 (see Fig. 1.2.6 for notation). The higher level gel allele is labelled as 1 (4.4 kb(A1)) and the lower one as 2 (3.9 kb(A2)). If we use this family for likeage analysis, what is the LOD score?*

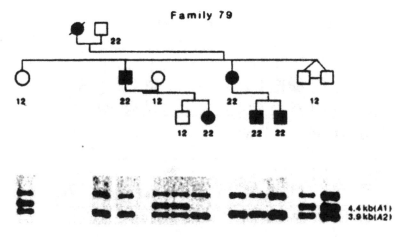

Figure 2.1.3: One family in Bodmer et al. (1987) for localizating the gene of FAP on chromosome 5. Location (probe) C11p11 has two alleles, $A1$ at 4.4 kb, $A2$ at 3.9 kb. (From Bodmer et al. (1987)), copyright ©by Macmillian Magazines Ltd. and the authors. Reprinted by permission.)

**Sol:** It can be easily seen that the second generation provides no recombination information. The deceased mother apparently had genotype 12 from the information of her offspring. No phase information can be obtained from her. Thus we have to assume equal probability for her genotype being $D1/d2$ or $D2/d1$, where $D$ means the dominant cancer gene. Since the second generation showed that there was either no recombination or all recombinations, the likelihood function for $\theta$ is

$$L(\theta) = \frac{1}{2}[\theta^4 + (1 - \theta)^4].$$

The maximum likelihood estimate of $\theta$ is $\theta = 0.0$, and LOD $= \log_{10}(2^3) = 0.903$. Of course, we cannot confirm the linkage with just one small family.         □

When six families, including the one in Fig. 2.1.3, are combined, the maximum LOD score is 3.26 at $\hat{\theta} = 0$, strong evidence of the linkage between C11p11 and FAP.

For a recessive disease gene, the computation is more complicated. We use the following example as a demonstration.

**Example. 2.1.5**  *Pras et al. (1994) used 17 families to locate the* cystinuria *gene. Cystinuria is an autosomal recessive disease characterized by excessive urinary excretion of cystine and the dibastic amino acids arginine, lysine and ornithine. The disease tends to form kidney stone in children as well as adults and may cause renal infection and failure. Approximately 1 in 60 Americans carries this recessive gene. By the position of the DNA responsible for the cystine, the authors concentrated their search on one chromosome. Fig. 2.1.4 gives one marker information for two families. The symbols for the microsatellites are from the upper gel to the lower gel 1, 2, 3 and 4.*

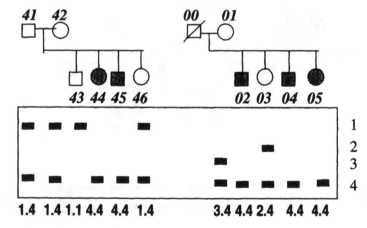

Figure 2.1.4: Two families in Pras et al. (1994) used to locate the *cystinuria* gene. The marker *D2S*391 has 4 alleles (*1, 2, 3, 4*) as shown in the gel photo, The numbers 41, 42 etc. are for personal identification and the symbols 1.4, 4.4, etc. are for alleles of *D2S*391.

Table 2.1.1: The probability of producing offspring 43–46 in Fig. 2.1.4.

| | | | Father $(D1/d4)$ | | | |
|---|---|---|---|---|---|---|
| | | Haplotype | D1 | d4 | D4 | d1 |
| | | Probability | $(1-\theta)/2$ | $(1-\theta)/2$ | $\theta/2$ | $\theta/2$ |
| | Hap. | Prob. | Notation | $f_1$ | $f_2$ | $f_3$ | $f_4$ |
| | D1 | $(1-\theta)/2$ | $m_1$ | 43 | 46 | 46 | 43 |
| Mother | d4 | $(1-\theta)/2$ | $m_2$ | 46 | 44, 45 | – | – |
| (D1/d4) | D4 | $\theta/2$ | $m_3$ | 46 | – | – | 46 |
| | d1 | $\theta/2$ | $m_4$ | 43 | – | 46 | – |

*The observed values are shown in gel and the numbers below the gel chart. What is the LOD score for the linkage between the disease gene and D2S391 in Fig. 2.1.4?*

**Sol:** Let $D$ again denote the dominant gene. This time, however, the disease is caused by the recessive $dd$ gene. For the family on the left, both parents have genotype $Dd$ 14. Without phase information we have to assume that there are two equal possible combinations $D1/d4$ or $D4/d1$ for each of them. For a given $\theta$, what is the likelihood that they will bear the genotypes of the offspring 43, 44, 45, 46? A special case with father $D1/d4$ and mother $D1/d4$ can be found in Table 2.1.1.

Note that the probability for an individual with this genotype is the sum of the products of the column and row probabilities it occupies. For example, the probability to produce 44 and 45 is $m_2 f_2 = (1-\theta)^2/4$, and to produce 43 is $(1-\theta)^2/4 + \theta(1-\theta)/2$ and to produce 46 is $(1-\theta)^2/2 + \theta(1-\theta)/2 + \theta^2/2$. Thus the likelihood of all four offspring is

$$L_1(\theta) = [(1-\theta)^2/4]^2[(1-\theta)^2/4 + \theta(1-\theta)/2][(1-\theta)^2/2 + \theta(1-\theta)/2 + \theta^2/2].$$

It can be shown that the likelihood function for $D1/d4$ and $D4/d1$ is simply to change the mother's haplotype probability from $(1-\theta)/2$ to $\theta/2$. We can find the other two possible combinations in a similar manner. However, it is better to do it systematically by introducing of function with the 8 haplotype probabilities as entries. Thus the likelihood becomes the product of 4 functions with,

$$L(\mathbf{f}, \mathbf{m}) = (m_2 f_2)^2[f_1 m_1 + f_1 m_4 + f_4 m_1][f_1(m_2 + m_3) + f_2 m_1 + f_3(m_1 + m_4) + f_4 m_3].$$
$$(2.1.6)$$

For the 01-05 family on the right, the genotype of the deceased person must be 2.4 from his offspring's 02 and 03. Thus, the genotype of the mother could be $D3/d4$ or $D4/d3$ and for the father could be $D2/d4$ or $D4/d2$. A table similar to Table 2.1.1 and a formula similar to (2.1.6) can be constructed. The computer output for the likelihood for the two families is shown in Fig. 2.1.5. A computer program for two generation families can be found in §B.1.

When all of the families are combined, the $Z_{max} = 3.73$ for microsatellite $D2S391$ with $\hat{\theta} = 0.15$. Another microsatellite $D2S119$, which is approximately $12\,\mathrm{cM}$ away from $D2S391$ has also been typed. It gave a very high LOD score 8.23 with $\hat{\theta} \approx 0$. This evidence should confirm that the *cystinuria* gene should in the neighborhood of $D2S119$.                                            □

Theoretically, an interval estimate for $\hat{\theta}$ can be obtained by analyzing the likelihood function using Fisher's information. But this is usually very complicated because, for any useful estimate of $\theta$, we either have a very large pedigree or many different small pedigrees. Conneally et al. (1985, see also Ott, 1991) considered a support domain of $\theta$ defined by $\{\theta : Z(\theta) \geq Z_{max} - 1\}$ as shown in Figs 2.1.5 and 2.1.6.

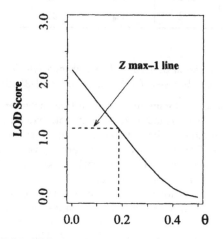

Figure 2.1.5: LOD $Z(\theta)$ and $Z_{max} - 1$ support interval for Fig. 2.1.4.

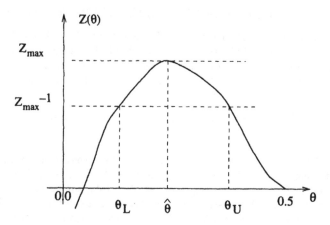

Figure 2.1.6: $Z_{max} - 1$ support interval $[\theta_L,\ \theta_U]$ for $\theta$.

This idea has since been widely adopted as a good confidence interval for $\theta$. We may reason it this way. The test defined by (2.1.4) does not have to test against $\theta = 0.5$. It can be extended to any $H_0: \theta = \theta_0$ against $H_1: \theta < \theta_0$ for a given $\theta_0 < 0.5$ and still have the mixed chi-square distribution (2.1.5). But when we move the $\theta_0$ close enough to $\hat{\theta}$, we can no longer claim them being significantly different. Thus, a good confidence interval for $\theta$ should cover this $\theta_0$, i.e., we should take all the $\theta_0$ as likely values if

$$Z_{\max} - Z(\theta_0) = \log_{10} \frac{Z(\hat{\theta})}{Z(\theta_0)} < c,$$

for some threshold $c$. Using Conneally et al.'s suggestion $c = 1$,

$$\Pr\left\{\log_{10} \frac{Z(\hat{\theta})}{Z(\theta_0)} < 1\right\} = 0.5 + 0.5\Pr\{\chi_1^2 < 4.6\} = 0.984.$$

Thus, the $Z_{\max} - 1$ is probably close to a 98.4% confidence interval. However, derivation of confidence interval by this reasoning is not rigorous. Ott (1991) says that we can claim this is at least a 90% confidence interval.

**Example. 2.1.6** *Larget-Piet et al. (1994) tried to locate the Usher syndrome gene. Usher syndrome, discovered by Usher in 1914, is an autosomal disorder characterized by congenital hearing impairment and progressive visual loss. It is regarded as a major cause of deaf-blindness in adults and accounts for 3 to 6% of deaf children. Fig. 2.1.7 shows the families they used. The detailed marker information was not given, but Fig. 2.1.8 shows the $Z_{\max}$ scores for 8 markers on chromosome 14 for a subset of the families that lived in the same region. The height $Z_{\max}$-score is 4.90 for marker D14S13 at $\hat{\theta} = 0$. The support interval $Z_{\max} - 1 = 3.90$ is shown on the x-axis. The authors also put other markers around D14S13 for a graphical representation of the gene's position.*

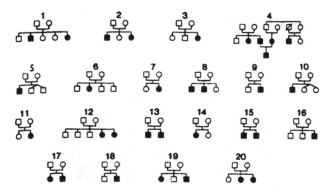

Figure 2.1.7: Families in Larget-Piet et al.'s Usher syndrome study. (From Larget-Piet et al. (1994), copyright ©by Academic Press, Inc. and the authors. Reprinted by permission.)

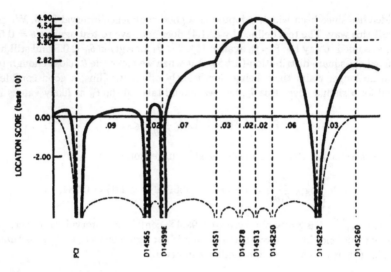

Figure 2.1.8: $Z$ scores in chromosome $14q$ in the neighborhood of markers $D14S13$. A $Z_{max} - 1$ support interval is shown by the horizontal dotted line. (From Larget-Piet et al. (1994), copyright ©by Academic Press, Inc. and the authors. Reprinted by permission.)

### 2.1.3   Sex-linked Disease Genes

To locate a sex-linked gene in general is easier than a non-sex-linked gene. We use Fig. 1.2.9 as an example. Since the gene is on the X-chromosome, the father's genetic information does not apply to male offspring. Let $d$ denote the recessive disease gene and $D$ the healthy gene. Since the female child $II1$'s genotype of the disease gene is unknown, she provides no information on recombination. The mother's genotype

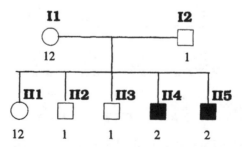

Figure 2.1.9: A pedigree for sex-linked disease. Note that there is only one marker for males.

should be $D1/d2$ and $D2/d1$ with probability 1/2 and 1/2. Thus, the likelihood equation is

$$L(\theta) = [\theta^4 + (1 - \theta)^4]/2.$$

Obviously, the female child should be considered if she has children who can reveal her genotype.

## 2.2 Linkage Analysis for Large Pedigrees

### 2.2.1 Theory

The concept of $\theta$ (recombination fraction) estimation for a large pedigree can be explained by the pedigree in Fig. 2.2.1. The goal is to find the recombination fraction $\theta$ between the disease gene and a marker, denoted by $A$ and $a$. In this example, we assume that the disease gene $D$ is autosomal and dominant and the affected persons are indicated by filled circles or squares. Let $d$ be the normal recessive allele.

Let the phenotype of individual $i$ be denoted by $x_i$. The individual in Fig. 2.2.1 is either affected or not affected. Let the genotype of the $i$th individual be denoted by $g_i$, which include all the possible genotypes. In this particular case, $g_i$ can be any one of the elements in the following set.

$$\{(DA/DA), (DA/Da), (Da/Da), (dA/dA), (da/dA), (dA/da), (DA/dA),$$
$$(Da/dA), (DA/da), (da/da)\}. \tag{2.2.1}$$

In general if there are $a_1$ alleles in locus 1 and $a_2$ alleles in locus 2, then the set (2.2.1) contains $k = a_1 a_2 (a_1 a_2 + 1)/2$ elements.

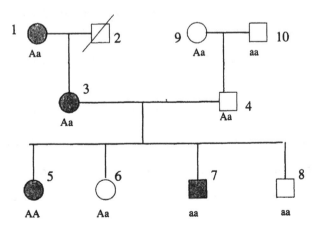

Figure 2.2.1: A pedigree with three generations. The marker has two alleles, $A$ and $a$.

Suppose there are $n$ individuals. Then the likelihood function

$$L(\theta) = P(x_1, x_2, \ldots, x_n | \theta)$$

$$= \sum_{g_1} \sum_{g_2} \cdots \sum_{g_n} P(x_1, x_2, \ldots, x_n | g_1, g_2, \ldots, g_n) P(g_1, g_2, \ldots, g_n) \qquad (2.2.2)$$

$$= \sum_{g_1} \sum_{g_2} \cdots \sum_{g_n} P(x_1 | g_1) P(x_2 | g_2) \ldots P(x_n | g_n) P(g_1, g_2, \ldots, g_n),$$

where $P(x_1, x_2, \ldots, x_n | g_1, g_2, \ldots, g_n) = \prod P(x_i | g_i)$ because of given genotypes, the phenotypes are independent. Some of the $\theta$s are omitted in the formulae for simplicity. This crude form is computationally impossible if a large number of summations need to be computed. The Elston-Stewart algorithm, later modified by Ott and implemented in *Liped* (see Appendix B.3), was to make ((2.2.2) computationally feasible.

Note that, in Fig. 2.2.1,

$$P(g_1, g_2, \ldots, g_n) = P(g_5, g_6, g_7, g_8 | g_1, g_2, g_3, g_4, g_9, g_{10}) P(g_1, g_2, g_3, g_4, g_9, g_{10}), \quad \text{and}$$

$$P(g_5, g_6, g_7, g_8 | g_1, g_2, g_3, g_4, g_9, g_{10}) = P(g_5, g_6, g_7, g_8 | g_3, g_4) = \prod_{i=5}^{8} P(g_i | g_3, g_4). \qquad (2.2.3)$$

We can now see how the pedigree information can be put into the likelihood. Suppose we have

$$g_3 = (DA/da), \quad g_4 = (dA/da), \quad \text{and} \quad g_5 = (DA/dA).$$

Then $P(x_5 | g_5) = 1$ and $P(g_5 | g_3, g_4) = (1 - \theta)/4$ (the product of $(1 - \theta)/2$ chance to get $DA$ from $g_3$ and $1/2$ chance to get $dA$ from $g_4$). On the other hand, if

$$g_3 = (DA/da), \quad g_4 = (da/da), \quad \text{and} \quad g_5 = (DA/dA),$$
$$\text{then} \quad P(x_5 | g_5) = 1 \quad \text{and} \quad P(g_5 | g_3, g_4) = 0; \quad \text{or if}$$
$$g_3 = (DA/da), \quad g_4 = (dA/da), \quad \text{and} \quad g_5 = (dA/dA),$$
$$\text{then} \quad P(x_5 | g_5) = 0 \quad \text{and} \quad P(g_5 | g_3, g_4) = \theta/4.$$

This is how the $\theta$ gets into the likelihood. The LOD score can be computed as follows:

$$\text{LOD}(\theta) = \log_{10} \left( \frac{L(\theta)}{L(\theta = 1/2)} \right).$$

We need now to compute $P(g_1, g_2, g_3, g_4, g_9, g_{10})$ in (2.2.3). One method is again to let

$$P(g_1, g_2, g_3, g_4, g_9, g_{10}) = P(g_3 | g_1, g_2, g_4, g_9, g_{10}) P(g_1, g_2, g_4, g_9, g_{10}), \qquad (2.2.4)$$

and $P(g_3 | g_1, g_2, g_4, g_9, g_{10}) = P(g_3 | g_1, g_2)$ can be easily computed. Again the term $P(g_1, g_2, g_4, g_9, g_{10})$ can be done by conditional probability

$$P(g_1, g_2, g_4, g_9, g_{10}) = P(g_4 | g_9, g_{10}) P(g_1, g_2, g_9, g_{10}). \qquad (2.2.5)$$

If we continue doing this recursively, eventually we end up computing the probabilities of the genotypes (with phase information) for 1, 2, 9, and 10. Actually (2.2.5) almost reaches that stage because

$$P(g_1, g_2, g_9, g_{10}) = P(g_1, g_2 | g_9, g_{10}) P(g_9, g_{10}) = P(g_1, g_2) P(g_9, g_{10}) \quad \text{and}$$

$$P(g_1, g_2) = P(g_1 | g_2) P(g_2) = P(g_1) P(g_2), \quad P(g_9, g_{10}) = P(g_9) P(g_{10}). \quad (2.2.6)$$

They will be computed by the population prior distribution. For example, for the deceased grandfather 2, a prior distribution at HWE will be assigned to all the 10 genotypes in the list (2.2.1). Of course, if D is a rare disease, some genotypes with D may be omitted due to their small likelihood. Similarly, for grandmother 1, a prior for

$$\{(DA/Da), (Da/dA), (DA/da)\}$$

has to be assigned by the population allele frequencies. Again for a rare disease, the chance of having genotype $(DA/Da)$ may be negligible.

Let us see the reduction of the computations. From (2.2.2), the number of computation is $O(k^{10})$, because $n = 10$ and each $g_i$ contains $k$ elements from (2.2.1). The first step (2.2.3) reduces the summation (2.2.2) to

$$\sum_{g_1} \sum_{g_2} \cdots \sum_{g_n} P(x_1 | g_1) P(x_2 | g_2) \ldots P(x_n | g_n) P(g_1, g_2, \ldots, g_n)$$

$$= \sum_{g_1} \cdots \sum_{g_4} \sum_{g_9} \sum_{g_{10}} P(x_1 | g_1) \ldots P(x_{10} | g_{10}) P(g_1, \ldots, g_4, g_9, g_{10})$$

$$\prod_{i=5}^{8} \left[ \sum_{g_i} P(x_i | g_i) P(g_i | g_3, g_4) \right]. \quad (2.2.7)$$

The computational complexity becomes $O(k^7)$. Though there is a reduction of a thousand fold in computation, $O(k^7)$ may still be a large number.

Elston and Stewart (1971) and Ott (1974) further reduced the computation by storing the likelihood values at certain pivotal points. For example, for given $g_3$ and $g_4$, the likelihood

$$L(5 - 8 | g_3, g_4) \equiv \prod_{i=5}^{8} \left[ \sum_{g_i} P(x_i | g_i) P(g_i | g_3, g_4) \right] \quad (2.2.8)$$

can be stored and reused. The computation of (2.2.8) is $O(k^3)$. Now if $g_3$ is fixed, then

$$L(4 - 10 | g_3) \equiv \sum_{g_4} \sum_{g_9} \sum_{g_{10}} P(x_4 | g_4) P(x_9 | g_9) P(x_{10} | g_{10}) L(5 - 8 | g_3, g_4)$$

$$P(g_4 | g_9, g_{10}) P(g_9) P(g_{10}) \quad (2.2.9)$$

This requires $O(k^4)$ computation (all combinations of $g_3, g_4, g_9, g_{10}$). When (2.2.9) is stored with index $g_3$, the rest of the computation is

$$\sum_{g_1} \sum_{g_2} \sum_{g_3} P(x_1 | g_1) P(x_2 | g_2) P(x_3 | g_3) L(4 - 10 | g_3)) P(g_3 | g_1, g_2) P(g_1) P(g_2),$$

with computation complexity $O(k^3)$. Thus the total computation for (2.2.7) is $O(k^3) + O(k^4) + O(k^3) = O(k^4)$.

Another way to reduce the computation without large storage is to use the Markovian property of a pedigree, i.e., conditioned on individuals 3 and 4, (1, 2) and (9, 10) are independent. Thus we may write (2.2.7) as

$$\sum_{g_3} P(x_3|g_3) \sum_{g_4} P(x_4|g_4) \left\{ \prod_{i=5}^{8} \left[ \sum_{g_i} P(x_i|g_i) P(g_i|g_3, g_4) \right] \right.$$

$$\cdot \left[ \sum_{g_1} \sum_{g_2} P(x_1|g_1) P(x_2|g_2) P(g_1, g_2|g_3) \right] \tag{2.2.10}$$

$$\left. \cdot \left[ \sum_{g_9} \sum_{g_{10}} P(x_9|g_9) P(x_{10}|g_{10}) P(g_9, g_{10}|g_4) \right] \right\}.$$

To compute $P(g_1, g_2|g_3)$, and similarly, $P(g_9, g_{10}|g_4)$, we use the Bayes' rule,

$$P(g_1, g_2|g_3) = \frac{P(g_3|g_1, g_2) P(g_1) P(g_2)}{\sum_{g_1} \sum_{g_2} P(g_3|g_1, g_2) P(g_1) P(g_2)}. \tag{2.2.11}$$

For given $g_3$, the computational complexity of the denominator of (2.2.11) is $O(k^2)$. However, the denominator of (2.2.11) needs only to be computed once for all the $g_1$ and $g_2$ in

$$\sum_{g_1} \sum_{g_2} P(x_1|g_1) P(x_2|g_2) P(g_1, g_2|g_3). \tag{2.2.12}$$

Thus, the computation of (2.2.12) is $O(k^2) + O(k^2) = O(k^2)$, and likelihood (2.2.10) can be computed again in $O(k^4)$.

**Example 2.2.1**  *Let us use a simple backcross dominant rare disease gene in Fig. 2.2.2 as an example.*

Since the gene is dominant rare, we can assume that individual 1 is $DA/da$ or $dA/Da$ and individual 2 is $da/da$ ($D$ stands for the disease gene and $d$ for the healthy one). A simple binomial distribution leads to

$$L(\theta) = \frac{1}{2} \left[ (1 - \theta)^5 \theta + \theta^5 (1 - \theta) \right]. \tag{2.2.13}$$

If (2.2.2) is used, the only non-zero terms are $\prod_{i=1}^{8} P(x_i|g_i) = 1$, which requires that all $P(x_i|g_i) = 1$. The requirement of $P(x_1|g_1) = P(x_2|g_2) = 1$ uses only the prior distributions of $g_1$ and $g_2$, i.e.,

$$P(g_1 = DA/da) = P(g_1 = dA/Da) = 0.5\pi_1, \quad \text{and} \quad P(g_2 = da/da) = \pi_2,$$

where $\pi_1$ and $\pi_2$ are the HWE probabilities for $Dd - Aa$ and $dd - aa$ respectively. In order for $\prod_{i=3}^{8} P(x_i|g_i) = 1$, it can be seen that

$$\prod_{i=3}^{8} P(g_i|g_1, g_2) = \begin{cases} (1 - \theta)^5 \theta & \text{if } g_1 = DA/da \\ (1 - \theta) \theta^5 & \text{if } g_1 = dA/Da \end{cases}$$

Thus, (2.2.2) is the same as (2.2.13) except a proportional constant.    □

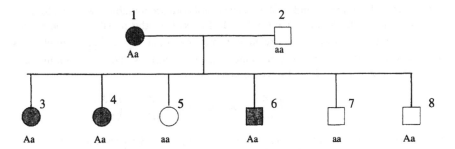

Figure 2.2.2: Pedigree for Example 2.2.1.

Suppose that the disease gene is recessive. Then the only choice for $g_1$ is $DA/Da$, and for $g_2$ is $Da/da$. No recombination information can be provided in Fig. 2.2.2. However, if the marker for $g_1$ were $aa$ and for $g_2$ were $Aa$, then we will get the same likelihood (2.2.2) for the linkage between the marker and the recessive gene.

Next we will see the necessity of using population allele frequency in linkage analysis.

**Example 2.2.2** *Find the LOD score for the recombination fraction between the two (co-dominant) markers for the pedigree in Fig. 2.2.3, the population frequency for 1 is 1/3, for 3 is 2/3, and for a and A are 0.4 and 0.6, respectively.*

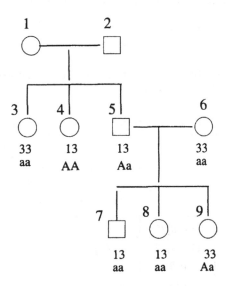

Figure 2.2.3: A pedigree including unknown genotypes.

**Sol**: Start with subjects 1 and 2. Since their genotypes are unknown, we have to use a population prior distribution like list (2.2.1) for each of the possible combinations. However, some of them can be ruled out. For example, $\binom{33}{aa} \times \binom{33}{AA}$ cannot be a possibility for persons 1 or 2. A careful check leads to only two possible combinations (symmetric cases in female and male do not count):

$$\text{Case 1: } \binom{13}{aA} \times \binom{13}{aA} \quad \text{or} \quad \text{Case 2: } \binom{13}{aA} \times \binom{33}{aA}.$$

For the given population frequencies $p_1 = 1/3$ and $p_3 = 2/3$ for the alleles $\{1, 3\}$ and under HWE, the prior probability for genotype (13) is $2p_1 p_3$ and for (33) is $p_3^2$. Since the $Aa$ genotypes are the same in both cases, the prior distribution of $a$ and $A$ will not affect the prior of the two cases. Consequently,

$$P(\text{Case 1}) : P(\text{Case 2}) = (2p_1 p_3)^2 : 2(2p_1 p_3) \times (p_3^2) = 1 : 2. \tag{2.2.14}$$

For case 1, there are three possible phase combinations:

$$\text{Ph1: } \binom{1|3}{a|A} \times \binom{1|3}{a|A}, \quad \text{Ph2: } \binom{1|3}{a|A} \times \binom{1|3}{A|a}, \quad \text{Ph3: } \binom{1|3}{A|a} \times \binom{1|3}{A|a}. \tag{2.2.15}$$

Their prior probabilities of occurrence are $P(\text{Ph1}) : P(\text{Ph2}) : P(\text{Ph3}) = 1 : 2 : 1$. If we use Ph1 as one example, the likelihood probability for Fig. 2.2.3 is now shown in Fig. 2.2.4. Note that individual 5 has two possible phases denoted by 5 and $5'$.

The details of the likelihood computation are given in Table 2.2.1. (See Table 2.1.1 for notation.)

Combined with the individuals 7, 8 and 9, the likelihood in Case 1 phase 1 becomes

$$L_{11} = [\theta^2/4][\theta(1-\theta)/2][(1-\theta)^2/2 \times (1-\theta)^3 + \theta^2/2 \times \theta^3]$$
$$= \theta^3(1-\theta)[(1-\theta)^5 + \theta^5]/16,$$

where in the first product, the first term is for individual 3, the second term for 4, and the third term for 5 and 7, 8, 9 and $5'$ and $7'$, $8'$ and $9'$.

Table 2.2.1: Likelihood computation for individuals 3, 4 and 5 for Fig. 2.2.4.

| | | | Father $\binom{1|3}{a|A}$ | | | |
|---|---|---|---|---|---|---|
| | | | $1a$ | $3A$ | $1A$ | $3a$ |
| | | | $(1-\theta)/2$ | $(1-\theta)/2$ | $\theta/2$ | $\theta/2$ |
| | $1a$ | $(1-\theta)/2$ | – | (5) | – | – |
| Mother | $3A$ | $(1-\theta)/2$ | (5) | – | (4) | – |
| $1a/3A$ | $1A$ | $\theta/2$ | – | (4) | – | $(5')$ |
| | $3a$ | $\theta/2$ | – | – | $(5')$ | (3) |

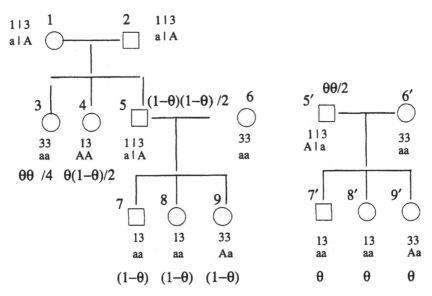

Figure 2.2.4: Likelihood for each individual for given phase information.

Table 2.2.2: Likelihood computation for individuals (3),
(4), (5) & (5') for $\left(\begin{smallmatrix}1|3\\a|A\end{smallmatrix}\right) \times \left(\begin{smallmatrix}3|3\\A|a\end{smallmatrix}\right)$.

|         |        |             | Father $\left(\begin{smallmatrix}3|3\\a|A\end{smallmatrix}\right)$ | |
|---------|--------|-------------|------|------|
|         |        |             | $3A$ | $3a$ |
|         |        |             | $1/2$ | $1/2$ |
|         | $1a$   | $(1-\theta)/2$ | (5)  | –    |
| Mother  | $3A$   | $(1-\theta)/2$ | –    | –    |
| $1a/3A$ | $1A$   | $\theta/2$  | (4)  | (5') |
|         | $3a$   | $\theta/2$  | –    | (3)  |

One example for Case 2 is given in Table 2.2.2; all the other cases are summarized in Table 2.2.3.

Hence, the overall likelihood for Fig. 2.2.3 is

$$L(\theta) = \frac{1}{4}L_{11} + \frac{2}{4}L_{12} + \frac{1}{4}L_{13} + 2\left[\frac{1}{2}L_{21} + \frac{1}{2}L_{22}\right]. \qquad (2.2.16)$$

In the above equation (2.2.16), the fractions 1/4, 2/4, 1/2 are the phase probabilities and the multiple 2 in the second two terms is due to (2.2.14).

The LOD score for this pedigree is found to be 0.116 at $\hat{\theta} = 0.2438$.                    □

Table 2.2.3: Likelihood computation for Fig. 2.2.3 with phase partitioning.

| Case (i) | Phase (j) | Individual likelihood for given $\theta$ | | | | Total ($L_{ij}$) |
|---|---|---|---|---|---|---|
| | | 3 | 4 | 5 | 5' | |
| 1 | 1: $\binom{1|3}{a|A} \times \binom{1|3}{a|A}$ | $\theta^2/4$ | $\theta(1-\theta)/2$ | $(1-\theta)^2/2$ | $\theta^2/2$ | $L_{11} = \theta^3(1-\theta)[(1-\theta)^5 + \theta^5]/16$ |
| 1 | 2: $\binom{1|3}{a|A} \times \binom{1|3}{A|a}$ | $\theta(1-\theta)/4$ | $[\theta^2 + (1-\theta)^2]/4$ | $\theta(1-\theta)/2$ | $\theta(1-\theta)/2$ | $L_{12} = \theta^2(1-\theta)^2[\theta^2 + (1-\theta)^2][(1-\theta)^3 + \theta^3]/32$ |
| 1 | 3: $\binom{1|3}{A|a} \times \binom{1|3}{A|a}$ | $(1-\theta)^2/4$ | $\theta(1-\theta)/2$ | $\theta^2/2$ | $(1-\theta)^2/2$ | $L_{13} = \theta^3(1-\theta)^5/16$ |
| 2 | 1: $\binom{1|3}{a|A} \times \binom{3|3}{A|a}$ | $\theta/4$ | $\theta/4$ | $(1-\theta)/4$ | $\theta/4$ | $L_{21} = \theta^2[(1-\theta)^4 + \theta^4]/64$ |
| 2 | 2: $\binom{1|3}{A|a} \times \binom{3|3}{A|a}$ | $(1-\theta)/4$ | $(1-\theta)/4$ | $\theta/4$ | $(1-\theta)/4$ | $L_{22} = \theta(1-\theta)^3[(1-\theta)^2 + \theta^2]/64$ |

Suppose the phenotype $x$ is quantitative with normal distribution $N(\mu_g, \sigma_g^2)$ for given genotype $g$. (The sigma is usually assumed to be the same for all genotypes.) Then

$$\prod_{i=1}^{n} P(x_i|g_i) = \prod_{i=1}^{n} \frac{1}{\sqrt{2\pi}\sigma_{g_i}} e^{-(x_i-\mu_{g_i})^2/(2\sigma_{g_i}^2)}. \qquad (2.2.17)$$

This is how (2.2.2) can be used for the quantitative trait. However, to maximize it by finding $\theta$ with unknown $\mu$s and $\sigma$s can be difficult. Some of the simpler cases will be discussed in Chapter 3.

## 2.2.2 Computational Tools

We can see that doing the pedigree by hand can be extremely time consuming. In Appendix B.1 a simple Fortran program to analysis two generation pedigree is included. Though its application in real data analyses is limited, it gives the basic concepts of linkage analysis.

One of the most popular computer packages for linkage analyses was developed by Ott's group. The procedures appear in Terwilliger and Ott (1994) and the computer programs can be obtained from their web site. Their book is more than 300 pages long, containing more than just pedigree analysis. The programs in their book are summarized in B.3. The following very large pedigree analysis was performed by an earlier version linkage program by Ott's group.

**Example 2.2.3** *Two pedigrees used by Watkins et al. (1986) to locate the cystic fibrosis (CF) gene are given in Fig. 2.2.5. (Cystic fibrosis is the most common autosomal recessive disorder in Caucasian children with a rate of 1 in 2000 children. Thus, the carrier proportion in Caucasian population is $\sqrt{1/2000} \approx 0.02$. It produces defective ion transport in the lung and pancreas and causes lung infection and pancreatic insufficiency. No effective treatment is known and it is generally fatal by the fourth decade.)*

Several markers were probed. One marker called *met* oncogene marker is used here for illustration. The results show that $\hat{\theta} = 0$ and $Z = 2.10$ for the Hutterite family and $\hat{\theta} = 0$ and $Z = 5.21$ for the Amish family. With combined z-score = 7.31, that there is linkage is without any doubt. When 17 other smaller families were combined with these, the total z-score became 15.45. Beaudet et al. (1986) combined Watkins et al. (1986) paper with several other papers and concluded that the CF gene must be in the neighborhood of marker *met* and D7S8 with the most likely order *met*-CF-D7S8 with recombination fractions $\theta(met, \text{CF}) = 0.01$ and $\theta(\text{CF}, \text{D7S8}) = 0.006$. The confidence limit was around 0.01 (1 cM).

## 2.2.3 Multi-point Analysis

Multi-point analysis tries to locate a gene using several adjacent marker loci simultaneously. Methods discussed so far, using one marker to locate a gene, are called

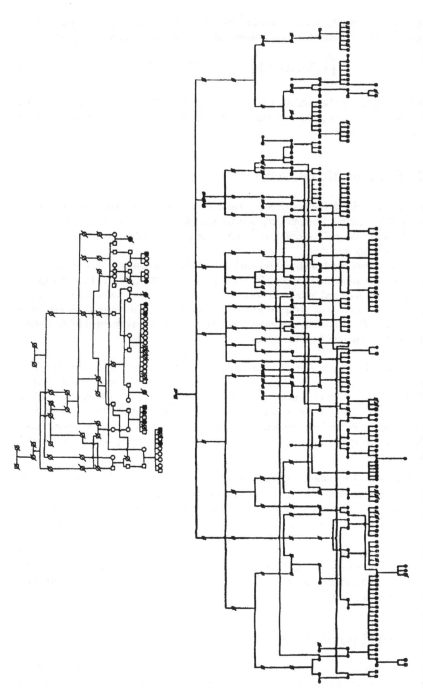

Figure 2.2.5: Two large pedigrees for finding the CF gene. The top figure is the Hutterite family and the bottom figure is the Amish-Mennonite family. (From Watkins et al. (1986), copyright ©by the University of Chicago Press. Reprinted by permission.)

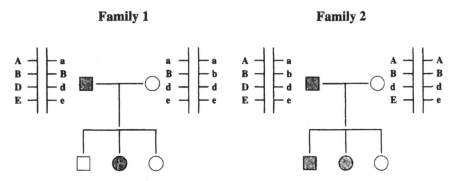

Figure 2.2.6: Two families with three marker loci A, B and E and a possible disease locus D.

two-point analyses. The advantage of multi-point analysis can be seen in Fig. 2.2.6, where $A, B, E$ are marker loci and $D$ is a possible disease gene locus. Assume that the disease allele $D$ is an autosomal dominant gene and $d$ is the recessive normal allele. If marker $B$ alone is used to locate this gene, then Family 1 in this figure provides no linkage information. However, if marker $A$ is added with a known distance between loci $A$ and $B$, then Family 1 can be useful in estimating the distance between $B$ and $D$ with data from other families such as Family 2. Obviously, the more markers the better, but the advantage of adding $A$ in the situation of Fig. 2.2.6 vanishes if $B$ is highly polymorphic. In this case, we should not have a $B/B$ genotype in any person. Even for highly polymorphic marker $B$, marker $E$ is still helpful to locate $D$ when $D$ is between $B$ and $E$. Since the position of $D$ is unknown, there are always merits in using multi-point analysis, even when all markers are highly polymorphic.

To quantify the merit of using additional markers can be very complicated. It depends on the inheritance mode (recessive, dominant, sex-linked), the polymorphism of the markers, phase information and the distances between the markers. We will use the following special case to illustrate the merit of using two flanking markers.

**Example 2.2.4.** *Suppose markers in loci $B$ and $E$ in Fig. 2.2.6 are highly polymorphic and $D$ is a rare dominant autosomal disease with two alleles $D$ (dominant) and $d$. Suppose we wish to test $H_0$ that $D$ is on a different chromosome against $H_1$ that $D$ is in the middle of loci $B$ and $E$. What is the advantage of using three-point analysis versus two-point analysis under the phase known condition? To quantify the advantage, define relative sample size gain as*

$$g(\theta) = \frac{n_2 - n_3}{n_2}, \qquad (2.2.18)$$

*where $n_i$ is the sample size needed to reach certain significance level $\alpha$ and power $1 - \beta$ in the i-point analysis and $\theta$ is the recombination fraction between $B$ and $D$ under $H_1$.*

**Sol**: Since $D$ is rare dominant and $B$ and $E$ are highly polymorphic, we may assume only one parent is informative. Let this parent's phenotype by $1D1/2d2$ (phase

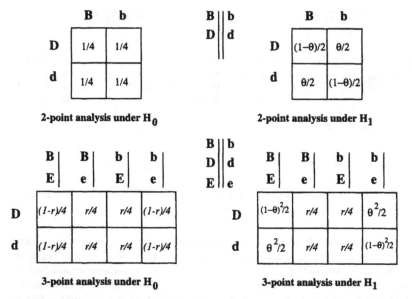

Figure 2.2.7: The probabilities of having different gametes after meiosis when one or two flanking markers are used. The disease alleles ($D$ and $d$) are supposed to be unlinked to $B$ and $E$ under $H_0$ and in the middle of $B$ and $E$ under $H_1$. The recombination fractions between $D$ and $B$ and $D$ and $E$ are both $\theta$ under $H_1$, and $r$ is the recombination fraction between $B$ and $E$.

known) under $H_1$ and the disease locus is on a different chromosome under $H_0$. The possible gametes are given in Fig. 2.2.7, and all of them are identifiable in the progeny. When one marker, marker $B$, is used, the two hypotheses are $H_0 : p = 1/2$ versus $H_1 : p = \theta$, where $p$ is the recombination fraction between locus $B$ and the disease locus. The sample size to satisfy significance level $\alpha$ and power $1 - \beta$ is

$$n_2(\theta) = \left[ \frac{z_\alpha \sqrt{p_0(1 - p_0)} + z_\beta \sqrt{p_1(1 - p_1)}}{p_0 - p_1} \right]^2,$$

by simple statistics, where $p_0 = 1/2$, $p_1 = \theta$ and $z_\alpha$ is the $\alpha 100\%$ upper quantile value of the standard normal distribution. When two markers are used, any offspring with gamete $Be$ or $bE$ is noninformative. Only progeny with $BE$ or $be$ can be used for linkage analysis. The two hypotheses are $H_0 : p = 1/2$ versus $H_1 : p = \theta^2/(1 - r)$, where $r = 2\theta(1 - \theta)$ and $p$ is the probability we see $BE - d$ or $be - D$ in the offspring who have markers $BE$ or $be$. The sample size requirement is

$$n_3(\theta) = \left[ \frac{z_\alpha \sqrt{p_0(1 - p_0)} + z_\beta \sqrt{p_1(1 - p_1)}}{p_0 - p_1} \right]^2 \bigg/ (1 - r),$$

where $p_0 = 1/2$ and $p_1 = \theta^2/(1 - r)$.

Numerical computation shows that the relative gains are always positive for $\theta > 0$ and if $\theta$ is not too small, the relative gain is in the neighborhood of 0.5 for small $\alpha$ and $\beta$. □

A similar discussion based on estimation error can be found in Lathrop et al. (1985).

To apply multi-point analysis, one has to order the markers and to estimate their mutual distances. When the number of markers is large and the marker loci are not so polymorphic, to make statistical inference on their order and distances is not simple. Interested readers may refer to Renwick and Bolling (1971) Cook et al. (1974), Pascoe and Morton (1987) and Zhao et al. (1990). Moreover, the estimated recombination fractions may cause contradiction (see Bailey, 1961 and Liberman and Karlin, 1984), and there may be interferences between the loci (see Exercise 1.8).

Suppose the markers are well order and their distances are accurately estimated. To use them to establish linkage can still be computation intractable using the approach in §2.2. The set of (2.2.1) may be too large. Possible genotypes increase exponentially as the number of marker loci increases. A program in Terwilliger and Ott (1994) can handle up to 5 marker loci simultaneously.

## 2.3 Incomplete Penetrance and Sibling Studies

So far the linkage analysis is based on the correct identification of the genotype from a given phenotype. For example, if a disease gene $d$ is recessive, then an unaffected individual must have genotype $Dd$ or $DD$ and an affected person must be $dd$. Though this is true for some genetic disease, such as sickle cell anemia, color blindness and muscular dystrophy, it is far from universal. Fig. 2.3.1 shows the expression of a breast cancer gene in various ages of women. The **penetrance**, defined as the conditional probability $P(phenotype|genotype)$, is age dependent. In this case, we can no longer say with certainty what the genotype is of an unaffected individual. When the phenotype can be clearly determined by the genotype, we call the gene's expression **full penetrance**, otherwise it is termed **incomplete penetrance**.

Note that in the likelihood function (2.2.2), only the term $P(x|g)$ depends on the penetrance. Let $A$ denote an affected person and $U$ denote an unaffected person. Let the onset rate given the existence of the disease genotype be $\lambda$. Then for the recessive disease case we have just discussed,

$$P(A|g) = \begin{cases} \lambda & \text{if} \quad g = dd \\ 0 & \text{if} \quad g \neq dd \end{cases}$$
$$P(U|g) = \begin{cases} 1 - \lambda & \text{if} \quad g = dd \\ 1 & \text{if} \quad g \neq dd, \end{cases} \tag{2.3.1}$$

and the likelihood function can still be constructed. If the probability of incomplete penetrance $\lambda$ is known, then linkage likelihood is still easy to maximize, although there

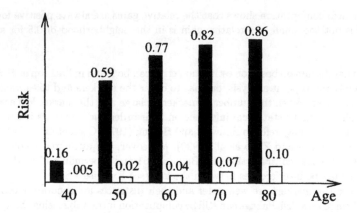

Figure 2.3.1: Age dependent risks of having breast cancer with a mutant breast-cancer gene (solid bars) and in the general population (unfilled bars).

is some loss efficiency in the estimation of $\theta$. To estimate how much efficiency is lost due to incomplete penetrance is important in sample size determination before data collection, but to compute the efficiency loss is not simple because it depends on the mating type. For example, if the mating type is $\binom{dD}{11} \times \binom{dd}{12}$, there is no linkage information. We will discuss this matter in detail in §2.4. Here we choose a **double backcross mating** $\binom{D|d}{2|1} \times \binom{d|d}{1|1}$ as an example, where D is the dominant healthy gene and the penetrance is defined by (2.3.1). Double backcross design is a commonly used design for linkage analysis in animal and plant gene search because the phases are known (see Fig. 2.3.2).

Let A11, A12, etc. represent an affected person with marker types 11, 12, etc. It can be seen in Fig. 2.3.2 that when $F_1 \times F_0(dd)$,

$$p_1 \equiv P(A11)$$
$$= \Pr\{\text{No crossover in } F_1, 1 \text{ is picked from} F_1, \text{subject affected}\}$$
$$+ \Pr\{\text{With crossover in} F_1, 1 \text{ is picked from} F_1, \text{subject affected}\}$$
$$= (1 - \theta)/2 \cdot \lambda + \theta/2 \cdot 0 = (1 - \theta)\lambda/2$$
$$p_2 \equiv P(A12) = \lambda\theta/2$$
$$p_3 \equiv P(U11) = (1 - \lambda + \lambda\theta)/2 \text{ and } p_4 \equiv P(U12) = (1 - \theta\lambda)/2.$$

Suppose there are, respectively, $n_1$, $n_2$, $n_3$, and $n_4$ A11, A12, U11, and U12 individuals in a sample of size $n = \Sigma n_i$. Then the Z-score at $\theta^*$ is

$$Z(\theta^*) = log_{10}(L(\theta^*)/L(0.5)),$$

with

$$L(\theta) = \prod_{i=1}^{4} p_i(\theta)^{n_i}.$$

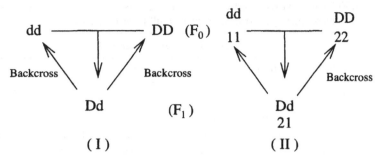

Figure 2.3.2: Backcross (I) ($Dd$ mates with $DD$ or $dd$) and double backcross (II) ($D2/d1$ mates with $DD/22$ or $dd/11$). The difference between (I) and (II) is in the number of loci.

Most likely, the estimated $\theta$, $\hat{\theta} \approx \theta$ and $n_i \approx np_i$. Thus, the expected LOD per person is

$$ELOD = \sum_{i=1}^{4} p_i \log_{10} \left( \frac{p_i(\theta)}{p_i(0.5)} \right). \qquad (2.3.2)$$

For example, when $\lambda = 0.5$ and $\theta = 0.1$,

$$
\begin{aligned}
ELOD = \frac{1}{2} \Big[ & (1 - 0.1) \times 0.5 \log_{10} 1.8 + 0.1 \times 0.5 \log_{10} 0.2 \\
& + (1 - 0.5 + 0.1 \times 0.5) \log_{10} \frac{1 - 0.5 + 0.1 \times 0.5}{1 - 0.5 \times 0.5} \\
& + (1 - 0.05) \log_{10} \frac{1 - 0.1 \times 0.5}{1 - 0.5 \times 0.5} \Big] = 0.0517.
\end{aligned}
$$

By constract, the full penetrance ELOD is 0.1595. Thus, the efficiency of the incomplete penetrance with $\lambda = 0.5$ is $0.0517/0.1595 = 0.323$. In other words, to reach the same LOD score for the linkage we need to increase the sample size by approximately three times. Table 2.3.1 gives the efficiency $\equiv$ ELOD (incomplete penetrance $\lambda$)/ ELOD (full penetrance $\lambda = 1$).

When $\lambda$ is unknown, we can still use the likelihood equation, but the parameters $\lambda$ as well as $\theta$ have to be estimated. When $\lambda$ is age-dependent and unknown, we have to group data according to the ages and deal with a large number of aged-dependent $\lambda$s. The estimates can then become very unreliable. This problem can be bypassed if we are only interested in testing for the existence of the linkage, but not in estimating the recombination fraction. The most commonly used method is the affected sibpair design.

## 2.3.1 Affected Sibpair Study

The initial idea of affected sibpair study by Penrose (1935) is very simple. First let us assume that the marker is very polymorphic so that the parents always have 4

Table 2.3.1: Efficiency of expected LOD scores at true recombination fraction, $\theta$, under incomplete penetrance, $\lambda$, versus full penetrance $\lambda = 1$.

| $\lambda$ | Recombination fraction, $\theta$ | | | | | |
|---|---|---|---|---|---|---|
| | 0.001 | 0.01 | 0.05 | 0.10 | 0.20 | 0.4990 |
| 1 | 1 | 1 | 1 | 1 | 1 | 1 |
| 0.95 | 0.857 | 0.867 | 0.883 | 0.891 | 0.899 | 0.905 |
| 0.90 | 0.760 | 0.769 | 0.787 | 0.798 | 0.809 | 0.818 |
| 0.80 | 0.611 | 0.618 | 0.634 | 0.644 | 0.656 | 0.667 |
| 0.50 | 0.312 | 0.314 | 0.319 | 0.323 | 0.328 | 0.333 |
| 0.25 | 0.138 | 0.139 | 0.140 | 0.141 | 0.142 | 0.143 |
| 0.10 | 0.052 | 0.052 | 0.052 | 0.052 | 0.053 | 0.053 |

different alleles $12 \times 34$ at the marker locus. Suppose the disease is genetic with incomplete penetrance, i.e., the gene is the necessary, but not the sufficient, condition for the disease. We are interested in testing:

$H_0$ : The marker and the gene are not linked ($\theta = 1/2$), against

$H_1$ : They are linked ($\theta < 1/2$).

Let $S_m$ denote the number of alleles shared by two affected siblings. (The subscript $m$ is for the marker, to be distinguished later from $t$, the trait alleles.) Then, under $H_0$,

$$P(S_m = 0) = 1/4, \quad P(S_m = 1) = 1/2, \quad P(S_m = 2) = 1/4. \qquad (2.3.3)$$

A chi-square test can be used to test the two hypotheses. Let there be $n$ sibpairs with $n_0$, $n_1$, and $n_2$ pairs sharing 0, 1 and 2 markers. Let $n = n_0 + n_1 + n_2$. Then,

$$Q = \sum_{i=0}^{2} \frac{[n_i - nP(S_m = i)]^2}{nP(S_m = i)} \qquad (2.3.4)$$

has chi-square with 2 degree of freedom (see A.4). However, the test base on (2.3.4) is not very specific for the alternative hypothesis $H_1$. For $H_1$ to be true, we expect a smaller $n_0$ and a larger $n_2$. Therefore, a more sensitive test is to test $n_2 - n_0$ directly on one side, i.e., reject $H_0$, if

$$Z = \frac{n_2 - n_0}{\sqrt{n/2}} \qquad (2.3.5)$$

is too large compared with a standard normal distribution. The derivation of the $z$ distribution is straightforward and left as an exercise.

For less polymorphic markers, the test becomes more complicated because (2.3.3) is no longer valid under the null hypothesis. For example, if the parents marker types are $aa \times ab$, then $P(S_m = 0) = 0$ is true under any linkage hypothesis. To proceed, we need the concept of **identity by descent** (*IBD*, *ibd*, or $\pi$), which is defined as two alleles being copies of a common ancestor. Since there is no way to trace one's remote ancestors, we will consider that two alleles are not *ibd* unless the two persons are related in recent generations. Thus, unless stated otherwise, we consider all matings random in a large population and the share of *ibd* between mates is 0. This is called **non-inbred** mating. Moreover, we assume that no mutation occurred in the short time of a pedigree study. Thus, two *ibd* alleles are identical, but the converse is not true. Two unrelated persons may have identical genotype at many loci. In non-inbred mating, identical twins have *ibd* = 2, parent and his/her offspring have *ibd* = 1, and two sibs have *ibd* = 0, 1, and 2 with probabilities 1/4, 1/2 and 1/4 respectively. Note that the average *ibd* between parent-offspring and sib-sib are the same, but the distributions are different. The advantage of using *ibd* is that it is independent of the distribution of the alleles. Note that in the case of $12 \times 34$ mating, $ibd = S_m$. However, in $11 \times 34$ mating, the distribution of $S_m$ changes, but not that of *ibd*.

Let $L_j(\theta)$ be the likelihood of having the observed markers from two affected sibs in family $j$. Then the joint likelihood function for $n$ families is

$$L(\theta) = \prod_{j=1}^{n} L_j(\theta). \tag{2.3.6}$$

Since parental markers are also important in the non-polymorphic case, we use the symbol $I_m$ to denote the marker information of the two sibs. For example $I_m(j) = (aa, ab)$ means that in the $j$th family, one sib has marker $(aa)$ and another has $(ab)$. Risch (1990c) and Holmans (1993) derived $L_j(\theta)$ as follows.

Let BSA denote the event that both sibs are affected and $\pi_m$ = the number of IBD marker alleles shared by two affected siblings. Then

$$
\begin{aligned}
L_j(\theta) &= P(I_m(j)|BSA) \\
&= \sum_{i=0}^{2} P(I_m(j)|\pi_m = i, BSA)P(\pi_m = i|BSA) \\
&= \sum_{i=0}^{2} P(I_m(j)|\pi_m = i)P(\pi_m = i|BSA) \\
&= \sum_{i=0}^{2} w_{ij}p_i,
\end{aligned}
\tag{2.3.7}
$$

where $w_{ij} = P(I_m(j)|\pi_m = i)$, and $p_i = P(\pi_m = i|BSA)$. Note that $p_i$ is independent of the family because $\pi_m$ depends only on the distance $\theta$ between the disease gene and the marker, but $w_{ij}$s are family dependent because they depend on the parents' marker type. It is apparent that under $H_0$ when the marker and the disease loci are not linked, $p_0 = p_2 = 1/4$ and $p_1 = 1/2$ as we have seen in (2.3.3) because at that time $S_m = \pi_m$ under the extreme polymorphic assumption. Moreover, $w_{ij}$s are either 0 or 1 and $L_j(\theta)$ is one of the $p_i$s. For a non-polymorphic case, $w_{ij}$ depends on the parental markers. For example, if the parental mating type is $aa \times ab$, and the two sibs have markers $(ab, ab)$, what are the $w_{ij}$s in (2.3.7)? One systematic way to find the $w_{ij}$s is to construct a table such as Table 2.3.2.

For $aa \times ab$, we have $1 = 2 = 3 = a$, and $4 = b$. Thus, the two sibs $(ab, ab)$ can only be the 4 cells $\{(14, 14), (14, 24), (24, 14), (24, 24)\}$ in Table 2.3.2. One can see that

Table 2.3.2: IBD number for the two sibs when the parents'
mating type is 12 × 34. The first column and the first row are
the marker types of the two sibs and the single digits 0, 1, or
2 in the lower right corner are their *IBD* numbers, $\pi_m$.

|  | sib 1 | | | |
|---|---|---|---|---|
| sib 2 | 13 | 14 | 23 | 24 |
| 13 | 2 | 1 | 1 | 0 |
| 14 | 1 | 2 | 0 | 1 |
| 23 | 1 | 0 | 2 | 1 |
| 24 | 0 | 1 | 1 | 2 |

$w_{0j} = P[ab, ab|\pi_m = 0] = 0/4 = 0, w_{1j} = P[ab, ab|\pi_m = 1) = 2/8 = 1/4$,   and   $w_{2j} = P[ab, ab|\pi_m = 2] = 2/4 = 1/2$. The reader can easily verify that if the parents' mating type is $ab \times ab$ with sib $I_m = (ab, ab)$, then $w_{oj} = w_{2j} = 1/2$, and $w_{1j} = 0$. Without parental marker information, the $w_{ij}$s are more difficult to compute because they depend on the marker allele frequencies. Let $PMT$ represent possible parental mating types. Then

$$P(I_m(j)|\pi_m = i)$$
$$= \sum_{PMT} P(I_m(j)|PMT \cap \pi_m = i)P(PMT|\pi_m = i)$$
$$= \sum_{PMT} P(I_m(j)|PMT \cap \pi_m = i)P(PMT), \tag{2.3.8}$$

where the summation includes all the possible parental mating types, and the last equation $P(PMT|\pi_m = i) = P(PMT)$ is because *ibd* is unrelated to the parental mating type. Computing (2.3.8) can be tedious, but tables are available in Risch (1990c). We again use the case $I_m = (ab, ab)$ as an example. Let the population frequency of $a$ be $f_1$, $b$ be $f_2$ and let $c$ denote any marker allele other than $a$ or $b$. Thus, $c$ may not be a single allele, but it has a total frequency $f_3 = 1 - f_1 - f_2$.

From $I_m = (ab, ab)$, the parents' mating type must be $ax \times by$, where $x, y$ can be $a, b$, and $c$. The relations between $PMT$, $P(PMT)$ and $w_{2j}(PMT) \equiv P(I_m(j) |PMT \cap \pi_m = 2)$ are given in Table 2.3.3. When summed up according to (2.3.8), $w_{2j} = \frac{1}{2}(4f_1^3 f_2) + 1 \cdot (2f_1^2 f_2^2) + \cdots = 2f_1 f_2$.

When $w_{ij}$ is computed for each family, the $L(\theta)$ in (2.3.6) contains only three unknowns, $p_0, p_1, p_2$, with $p_0 + p_1 + p_2 = 1$. We know that under $H_0 : p_1 = 1/2$ and $p_0 = p_2 = 1/4$ and under $H_1 : p_2$ is higher than $p_1$. Thus, we can use the likelihood ratio test

$$T = \log_{10}[L(\hat{p}_0, \hat{p}_1, \hat{p}_2)/L(H_0)], \tag{2.3.9}$$

where $\hat{p}_i$s are the maximum of $L(p_0, p_1, p_2)$ by varying $p_0, p_1$, and $p_2$ and $L(H_0)$ is the value in (2.3.6) when $p_0 = p_2 = 1/4$, $p_1 = 1/2$. Since we cannot directly maximize (2.3.6) by $\theta$, we cannot estimate $\theta$ from (2.3.6). The original likelihood function on $\theta$ in

Table 2.3.3: Computing (2.3.8) by listing all the feasible parental mating types *PMT*.

| *PMT* | $aa \times ab$ | $aa \times bb$ | $aa \times bc$ | $ab \times ba$ | $ab \times bb$ |
|---|---|---|---|---|---|
| $P(PMT)$ | $4f_1^3 f_2$ | $2f_1^2 f_2^2$ | $4f_1^2 f_2 f_3$ | $4f_1^2 f_2^2$ | $4f_1 f_2^3$ |
| $w_{2j}(PMT)$ | $1/2$ | $1$ | $1/2$ | $1/2$ | $1/2$ |

| *PMT* | $ab \times bc$ | $ac \times ba$ | $ac \times bb$ | $ac \times bc$ |
|---|---|---|---|---|
| $P(PMT)$ | $8f_1 f_2^2 f_3$ | $8f_1^2 f_2 f_3$ | $4f_1 f_2^2 f_3$ | $8f_1 f_2 f_3^2$ |
| $w_{2j}(PMT)$ | $1/4$ | $1/4$ | $1/2$ | $1/4$ |

$P(PMT)$ denotes the PMT frequency under *HWE*, and $w_{2j}(PMT) = P(I_m(j) \,|\, PMT \cap \pi_m) = 2$, where $j$ is the family index.

(2.3.6) has been replaced by the likelihood on $(p_0, p_1, p_2)$. To estimate $\theta$, we have to find the $P(\pi_m = i | BSA)$ in terms of $\theta$. We will do this later in this section. According to the likelihood ratio test, we should have $4.6T \sim \chi_2^2$, but that is not true due to the boundary conditions we have discussed in 2.1. Detailed discussion can be found in Holmans (1993). Usually, $T \geq 3$ is considered a good rule to claim linkage (see Risch, 1990c).

It can be seen that $L_j(\theta)$ in (2.3.7) can be extended to any affected pairs, whether they are sibs, cousins, or uncle/nephews. Details can be found in Risch (1990c).

## 2.3.2   Affected Sib Study with More Than Two Affected Siblings

An obvious inconvenience of the affected sibpair study is that only one pair of siblings is allowed. What happens if there are more than two affected siblings? A quick solution is to discard some of them randomly, as suggested by Thomson (1980). Or, we may use all the possible pairs as suggested by Weitkamp et al. (1981), but the dependence makes the standard distributions of (2.3.4) and (2.3.5) invalid. Green and his colleagues (Green et al., 1983, Green and Low, 1984) devised an **N-test** to handle this situation. The N-test is also equivalent to the F-test suggested by De Vries et al. (1976).

Suppose the marker is fully polymorphic in the two parents. Let there be $k$ families and the number of affected siblings be $s_i$ ($s_i \geq 2$) in the $i$th family, $i = 1, 2, \ldots, k$. They define

$N_i$ = the sum over the most frequently occurring haplotypes, one from each parent.

Table 2.3.4 shows an example on how this definition applies.

The N-test statistic is

$$N = \sum_{i=1}^{k} N_i. \tag{2.3.10}$$

Table 2.3.4: Definition of $N_i$ with 4 affected siblings with parental mating type $ab \times cd$. Symmetric cases such as $(ad, ad, ad, ad)$ are omitted.

|           | ac | ac | ac | ac | ac | ac |
|-----------|----|----|----|----|----|----|
| Sib       | ac | ac | ac | ac | ac | ac |
| Haplotypes| ac | ac | ac | ad | ad | bd |
|           | ac | ad | bd | ad | bd | bd |
| $N_i$     | 8  | 7  | 6  | 6  | 5  | 4  |

Since the families are independent, so $N$ should have an approximate normal distribution for large $k$. Thus we only need to find out $E(N_i)$ and $V(N_i)$. For the $s_i$ affected siblings, let $K_i(m)$ denote the number of one of the alleles in the sibs from the mother side. We define $K_i(f)$ from the father's side similarly. Then under $H_0$, $K_i(m) \sim \text{Bi}(s_i, 1/2)$ and the number of more frequently occurring marker $N_i(m) \equiv \max(K_i(m), s_i - K_i(m))$. The distribution of $N_i(m)$ is found to be

$$Pr\{N_i(m) = j\} = \begin{cases} 2\binom{s_i}{j}(1/2)^{s_i} & \text{for all } j > s_i/2 \\ \binom{s_i}{j}(1/2)^{s_i} & \text{for } j = s_i/2, \text{ when } s_i \text{ is even.} \end{cases}$$

The computation of $E(N_i(m))$ and $V(N_i(m))$ is simple by a computer. Green and Low (1984) worked out formulae for the $E(N_i(m))$ and $V(N_i(m))$ and listed their values in Table 2.3.5. Since $N_i = N_i(m) + N_i(f)$ and $N_i(m)$ and $N_i(f)$ are independent and identically distributed, $E(N_i) = 2E(N_i(m))$ and $V(N_i) = 2V(N_i(m))$ under $H_0$.

**Example 2.3.1** *The human immune system has a large variety of alleles in order to produce different immune proteins. (That is why organ transplantation is usually difficult because the grafted organ may be rejected by the host's immune system.) A group of important immune genes, **human leukocyte antigen (HLA)**, HLA-A, HLA-B, HLA-C, HLA-D, are located tightly together at 6q21. Their combined variations, more than 100 in total, can be considered as fully polymorphic. They are easily*

Table 2.3.5: Mean and variance of $N_i$ (subscript $i$ dropped in the table).

| $s$ | $E(N)$ | $V(N)$ |
|-----|--------|--------|
| 2   | 3.0    | 0.5    |
| 3   | 4.5    | 0.375  |
| 4   | 5.5    | 0.875  |
| 5   | 6.875  | 0.7422 |
| 6   | 7.875  | 1.2422 |
| 7   | 9.1875 | 1.1074 |
| 8   | 10.1875| 1.6074 |

Table 2.3.6: Cudworth's diabetes data. (*Used by permission of the International Biometric Society.*)

| $s$ | $N_i$ | | | | | | | Total # of families |
|---|---|---|---|---|---|---|---|---|
| | 2 | 3 | 4 | 5 | 6 | 7 | 8 | |
| 2 | 7 | 46 | 69 | | | | | 122 |
| 3 | | | 2 | 4 | 4 | | | 10 |
| 4 | | | | | | 1 | | 1 |
| Total | 7 | 46 | 71 | 4 | 4 | 1 | | 133 |

detectable in blood tests at the protein level. Hence they provided a convenient tool for linkage study before DNA type of polymorphism were available. Green et al. (1983, 1984) used HLA to test the linkage between HLA and an insulin-dependent diabetes (IDDM) gene using affected siblings in 131 families. The data is given in Table 2.3.6. What statistical inference can we make regarding the linkage between IDDM and HLA?

**Sol**: By (2.3.10),

$$N = 2 \times 7 + 3 \times 46 + 4 \times 71 + 5 \times 4 + 6 \times 4 + 8 \times 1 = 488.$$

Under $H_0$ and using Table 2.3.5, $E(N) = 122 \times 3 + 4.5 \times 10 + 5.5 = 416.5$, and $V(N) = 122 \times 0.5 + 10 \times 0.375 + 0.875 = 65.625$. Thus, the test statistic by the central limit theorem is

$$Z = \frac{488 - 416.5}{\sqrt{65.625}} = 8.826.$$

There is little doubt that the data showed the linkage between HLA and a diabetes gene. □

**Remark**: The test significance of this data does not imply that all the diabetes genes are located in the HLA region. It is quite possible there are other diabetes genes located in other places. This is similar to testing the mean $\mu = 0$ under $H_0$, when the population is a mixture of two normals $N(0, \sigma^2)$ and $N(1, \sigma^2)$. A sample from this population will reject $H_0$, but it cannot guarantee there is no subpopulation with mean 0. A subsequent study by Davies et al. (1994) shows that there are many sources for diabetes; IDDM4 is linked to 11q13. The study by Davies et al. was also based on sibpairs but used Risch's (1990a, b, c) method, which will be discussed later in this section. More diabetes genes have been found in 6q25, 6q27 and 2q33 (see Marron et al., 1997).

Green and Low's method can be extended to not-so-polymorphic markers if the parents' markers are known. For example. if the parents are *12 × 12* for the *i*th family, then the distribution of $N_i$ for 2 affected sibs under $H_0$ can be listed. Any offspring with markers *11, 12, 21,* or *22* (*12* and *21* are indistinguishable.) has probability 1/4. Thus,

$$P(N_i = 4) = 1/4, \quad P(N_i = 3) = 1/2, \quad \text{and} \quad P(N_i = 2) = 1/4,$$

with the convention that two siblings with *12* and *12* will be considered as $N = 4$ and $N = 2$ with equal chance.

Green and Grennan (1991) further extended their results to cases with unknown parental markers. Their method was to infer the parental marker type from the alleles of all the siblings. With the number of siblings and the prior distribution of the markers, we can find the posterior distribution of the parents' mating type, although many tables are necessary for all the possible combinations. The tables can be found in Green and Grennan's paper. Instead, a similar method by Lange (1986) is presented here.

Suppose that in a family there are $s$ affected siblings. Define

$$Z = \sum_{i=1}^{s} \sum_{j=i+1}^{s} X_{ij}, \quad X_{ij} = \text{the number of shared alleles by individuals } i \text{ and } j.$$

To test $H_0$, we use the sum of a weighted sum of the $Z$s over all the families. By the central limit theorem, we need only to compute the mean and variance of $Z$. Thus,

$$E[Z] = \binom{s}{2} E[X_{ij}], \quad \text{and}$$

$$E[Z^2] = \sum_{i<j} \sum_{k<\ell} E[X_{ij} X_{k\ell}]$$

$$= \binom{s}{2} E[X_{ij}^2] + s(s-1)(s-2) E[X_{ij} X_{i\ell}] + \binom{s}{2} \binom{s-2}{2} E[X_{ij} X_{k\ell}],$$

where different symbols are not equal, e.g., $j \neq \ell$, $k \neq i$. The values $E[X_{ij}]$, $E[X_{ij} X_{i\ell}]$, etc. depend on the parents' mating types. Lange suggests that we use their population frequencies under HWE. Let $p_m$ be the population frequency for allele $m$. Table 2.3.7 gives the distribution of the mating types.

Table 2.3.7: Distribution of mating types under HWE.

| $M(Mating\ Type)$ | Probability |
|---|---|
| $aa \times aa$ | $\sum_m p_m^4$ |
| $aa \times bb$ | $\left(\sum_m p_m^2\right)^2 - \sum_m p_m^4$ |
| $aa \times ab$ | $\sum_m 4p_m^3(1 - p_m)$ |
| $aa \times bc$ | $\sum_m 2p_m^2(1 - \sum_n p_n^2 - 2p_m(1 - p_m))$ |
| $ab \times ab$ | $\sum_{m<n}(2p_m p_n)^2$ |
| $ab \times ac$ | $\sum_{m<n} 4p_m p_n(p_m + p_n)(1 - p_m - p_n)$ |
| $ab \times cd$ | $\sum_{m<n} 2p_m p_n(1 - 2p_m p_n - 2(p_m + p_n)(1 - p_m - p_n) - \sum_k p_k^2).$ |

In the table, $p_m$ or $p_n$ are population frequencies for alleles $m$ or $n$. The summation are over all the alleles. The letters $a$, $b$, $c$ and $d$ represent any 4 possible alleles.

Table 2.3.8: Computation of conditional expectations (adapted from Lange,1986).

| $M$ (Mating type) | $E[X_{ij}|M]$ | $E[X_{ij}X_{ij}|M]$ | $E[X_{ij}X_{i\ell}|M]$ | $E[X_{ij}X_{k\ell}|M]$ |
|---|---|---|---|---|
| $aa \times aa$ | 2 | 4 | 4 | 4 |
| $aa \times bb$ | 2 | 4 | 4 | 4 |
| $aa \times ab$ | 3/2 | 5/2 | 9/4 | 9/4 |
| $aa \times bc$ | 3/2 | 5/2 | 9/4 | 9/4 |
| $ab \times ab$ | 5/4 | 2 | 13/8 | 25/16 |
| $ab \times ac$ | 9/8 | 13/8 | 41/32 | 81/32 |
| $ab \times cd$ | 1 | 3/2 | 1 | 1 |

Table 2.3.9: Computation of $E[X_{ij}X_{ij}|M = ab \times ac]$
(adapted from Lange, 1986).

| Sib genotype | | Value of $X_{ij}X_{ij}$ | Sib genotype | | Value of $X_{ij}X_{ij}$ |
|---|---|---|---|---|---|
| $i$ | $j$ | | $i$ | $j$ | |
| aa | aa | 4 | ab | aa | 1 |
| aa | ac | 1 | ab | ac | 1 |
| aa | ab | 1 | ab | ab | 4 |
| aa | bc | 0 | ab | bc | 1 |
| ac | aa | 1 | bc | aa | 0 |
| ac | ac | 4 | bc | ac | 1 |
| ac | ab | 1 | bc | ab | 1 |
| ac | bc | 1 | bc | bc | 4 |
| Total | | | | | 26 |

Given mating type $M$, the conditional expections $E[X_{ij}|M]$, $E[X_{ij}X_{i\ell}|M]$, etc. can be found in Table 2.3.8, and the derivation of one example is given in Table 2.3.9.

Thus, $E[X_{ij}X_{ij}|M = ab \times ac] = 26/16 = 13/8 =$ the value in Table 2.3.8.

The mean of $X_{ij}$ can be computed by

$$E[X_{ij}] = \sum_t E[X_{ij}|M = t]P[M = t],$$

using Tables 2.3.7 and 2.3.8. The variance for $Z$, $\sigma_s^2 = E[Z^2] - E[Z]^2$ can also be computed. Due to possible large combinatoric numbers $\binom{s}{2}$, the sum of the $Z$s would be biased toward a large family. Thus, a reasonable statistic for testing $H_0$ is to use the minimum variance sum

$$T = \sum_{i=1}^{k} Z_i/\sigma_{s_i}^2,$$

where $k$ is the total number of families in the study and $i$ is the family index.

### 2.3.3   Using Affected Sibpairs to Locate the Gene

The test for linkage by affected siblings cannot be used to estimate the distance between the disease and marker loci. Also, the unaffected siblings are not used, because their trait genotypes are uncertain. The disease genotype depend on the onset rate. Moreover, the parents' trait genotypes cannot be confidently defined either. Thus, it is not possible to estimate the recombination fraction $\theta$ in incomplete penetrance unless we can estimate the onset rate $\lambda$. As stated before, we may combine $\theta$ and $\lambda$ as unknowns in the likelihood equation. However, in doing this all the siblings or other relatives in the pedigree have to be DNA typed. This may not be cost effective because an unaffected person usually provides little linkage information. Risch (1990 a, b, c) in a series of papers advocates that only affected siblings or relatives be used in a likelihood equation. All the other relatives can be used to estimate the onset rates (disease risks). He defines $\lambda$ as follows:

$\lambda_p$ = population disease risk,

$\lambda_o$ = risk of a offspring, given that one of the parents is affected,

$\lambda_s$ = risk of a sibling, given that one child is affected,

$\lambda_m$ = risk of a monozygotic twin, given that one of them is affected.

For a large population survey, we should be able to estimate the first three risks with great accuracy. There may not be enough monozygotic twins with a rare disease to estimate $\lambda_m$, but it can be estimated using the formula

$$\lambda_m = 4\lambda_s - 2\lambda_o - \lambda_p. \tag{2.3.11}$$

Since the proof of (2.3.11) involves some new concepts in genetics, it is left until the end of this section.

Go back to (2.3.6), we can put $\theta$ into the likelihood equation if we can put $\theta$ into

$$p_i = P(\pi_m = i | BSA). \tag{2.3.12}$$

in (2.3.7). By the Bayes theorem,

$$p_i = P(\pi_m = i | \text{BSA}) = \frac{P(\text{BSA} | \pi_m = i) p_o(\pi_m = i)}{P(\text{BSA})}, \tag{2.3.13}$$

where BSA = both sibs are affected as before, $p_o(\pi_m = i)$ is the prior distribution, which can be found from (2.3.3). The denominator of (2.3.13) is $\sum_i P(BSA|\pi_m = i)$ $p_o(\pi_m = i)$ once we know how to compute $P(BSA|\pi_m = i)$ or simply let $P(BSA) = \lambda_s \lambda_p$.

Let $\pi_t$ and $\pi_m$ be the IBD numbers shared by the two sibs at the trait and the marker loci respectively. Thus, the first term in the numerator of (2.3.13) becomes

$$P(\text{BSA}|\pi_m = i) = \sum_{j=0}^{2} P(\text{BSA}|\pi_t = j)P(\pi_t = j|\pi_m = i). \tag{2.3.14}$$

Table 2.3.10 Values of $P(\pi_t = j | \pi_m = i)$.

| | | $\pi_t$ | | |
| | | 2 | 1 | 0 |
|---|---|---|---|---|
| | 2 | $\Psi^2$ | $2\Psi(1 - \psi)$ | $(1 - \Psi)^2$ |
| $\pi_m$ | 1 | $\Psi(1 - \Psi)$ | $\Psi^2 + (1 - \Psi)^2$ | $\Psi(1 - \Psi)$ |
| | 0 | $(1 - \Psi)^2$ | $2\Psi(1 - \Psi)$ | $\Psi^2$ |

The last term is the term that related to $\theta$. If we let $\Psi = \theta^2 + (1 - \theta)^2$ as defined by Haseman and Elston (1972), the conditional probabilities $P(\pi_t = j | \pi_m = i)$ are given in Table 2.3.10.

We use $i = 2$, $j = 2$ as an example.

**Example 2.3.2** *Show that* $P(\pi_t = 2 | \pi_m = 2) = \Psi^2$.

**Sol**: The segregation of alleles in two loci can be expressed as shown in Fig. 2.3.3. From Fig. 2.3.3, we have,

$$P(\pi_t = \pi_m = 2) = P(\text{Two individuals have the same diploids})$$

$$= \left[ \sum_{i=1}^{4} P(\text{2nd person} = f_i | \text{1st person} = f_i) P(\text{1st person} = f_i) \right]$$

$$\times \left[ \sum_{i=1}^{4} P(\text{2nd person} = m_i | \text{1st person} = m_i) P(\text{1st person} = m_i) \right]$$

$$= \left[ \left( \frac{1 - \theta}{2} \right)^2 \times 2 + \left( \frac{\theta}{2} \right)^2 \times 2 \right]^2 = \frac{\Psi^2}{4}.$$

By $P(\pi_t = j | \pi_m = i) = P(\pi_t = \pi_m = 2) / P(\pi_m = 2)$ and $P(\pi_m = 2) = 1/4$, we have the value in Table 2.3.10. $\square$

| Parent | | $M_1 -\vert$ | $\vert\!- M_2$ | | | $M_3 -\vert$ | $\vert\!- M_4$ | |
| type | | $T_1\vert$ | $\vert\!- T_2$ | | | $T_3 -\vert$ | $\vert\!- T_4$ | |
| | | $\downarrow$ | | | | $\downarrow$ | | |
| Haplo- | $M_1 -\vert$ | $M_2 -\vert$ | $M_1 -\vert$ | $M_2 -\vert$ | $\vert\!- M_3$ | $\vert\!- M_4$ | $\vert\!- M_3$ | $\vert\!- M_4$ |
| type | $T_1 -\vert$ | $T_2 -\vert$ | $T_2 -\vert$ | $T_1 -\vert$ | $\vert\!- T_3$ | $\vert\!- T_4$ | $\vert\!- T_4$ | $\vert\!- T_3$ |
| Symbol | $f_1$ | $f_2$ | $f_3$ | $f_4$ | $m_1$ | $m_2$ | $m_3$ | $m_4$ |
| Prob. | $(1 - \theta)/2$ | $(1 - \theta)/2$ | $\theta/2$ | $\theta/2$ | $(1 - \theta)/2$ | $(1 - \theta)/2$ | $\theta/2$ | $\theta/2$ |

Figure 2.3.3: Segregation of two polymorphic parents.

Now, the first term in the summation of (2.3.14) can be computed by

$$P(BSA|\pi_t = j) = \begin{cases} \text{P(Two unrelated persons affd.)} & = \lambda_p^2 & \text{for } j = 0 \\ \text{P(Parent-offspring both affd.)} & = \lambda_p\lambda_o & \text{for } j = 1 \\ \text{P(Two monozygotic twins affd.)} = \lambda_p\lambda_m & \text{for } j = 2. \end{cases} \quad (2.3.15)$$

Combining (2.3.13)–(2.3.15), we have,

$$\begin{aligned} p_0 &= \frac{1}{4} - \frac{1}{4\lambda_s}(2\Psi - 1)[(\lambda_s - \lambda_p) + 2(1 - \Psi)(\lambda_s - \lambda_o)] \\ p_1 &= \frac{1}{2} - \frac{1}{2}(2\Psi - 1)^2(1 - \lambda_o/\lambda_s) \qquad (2.3.16) \\ p_2 &= \frac{1}{4} + \frac{1}{4\lambda_s}(2\Psi - 1)[(\lambda_s - \lambda_p) + 2\Psi(\lambda_s - \lambda_o)] \end{aligned}$$

It takes some algebraic manipulation to arrive at the three $p$s in these forms. For the special case $\theta = 1/2$, or $\Psi = 1/2$, the $p$'s are 1/4, 1/2, and 1/4.

When the $\lambda$ in (2.3.15) are reliably estimated, the likelihood equation (2.3.12) involves only one parameter $\theta$ and can be easily maximized. Note also since every term in (2.3.15) has a factor $\lambda_p$, we may omit one $\lambda_p$ from each of the probabilities without affecting the maximum likelihood procedure.

To prove (2.3.11), define a random variable $X$ for an individual such that $X = 1$ if the person is affected and 0 otherwise. Similarly, define $Y$ for the other person. Obviously, $E[X] = E[Y] = \lambda_p$. Let Cov(X, Y) denote the covariance between the two individuals. Then

$$\text{Cov}(X,Y) = E[(X - \lambda_p)(Y - \lambda_p)] = E[XY] - \lambda_p^2 = P(X = Y = 1) - \lambda_p^2.$$

This formula implies the result that Cov(X, Y) = 0 if the two persons are unrelated. Let $IBD$ denote the $IBD$ number between $X$ and $Y$. Using conditional probability, we have

$$\text{Cov}(X,Y) = \sum_{i=0}^{2} E[(X - \lambda_p)(Y - \lambda_p)|IBD = i]P(IBD = i)$$

$$= \sum_{i=1}^{2} E[(X - \lambda_p)(Y - \lambda_p)|IBD = i]P(IBD = i).$$

To simplify the notation, let $V_1 = E[(X - \lambda_p)(Y - \lambda_p)|IBD = 1]$ and $V_2 = E[(X - \lambda_p)(Y - \lambda_p)|IBD = 2]$. Then we have

$$\text{Cov}(X,Y) = V_1 P(IBD = 1) + V_2 P(IBD = 2). \qquad (2.3.17)$$

On the other hand,

$$\begin{aligned} \text{Cov}(X,Y) &= P(X = Y = 1) - \lambda_p^2 \\ &= Pr\{\text{Both X and Y are affected}\} - \lambda_p^2 \\ &= Pr\{\text{X is affected}|\text{Y is affected}\}Pr\{\text{Y is affected}\} - \lambda_p^2 \\ &= \lambda_R\lambda_p - \lambda_p^2, \end{aligned}$$

implies

$$V_1 P(IBD = 1) + V_2 P(IBD = 2) = \lambda_R \lambda_p - \lambda_p^2, \qquad (2.3.18)$$

where the relationship between X and Y is denoted by $R$, which can be one of the $m$, $o$ or $s$ in (2.3.11). Substituting $m$, $o$ and $s$ for $R$ in (2.3.18), we have three equations:

$$V_2 = \lambda_m \lambda_p - \lambda_p^2,$$
$$V_1 = \lambda_o \lambda_p - \lambda_p^2,$$
$$V_1/2 + V_2/4 = \lambda_s \lambda_p - \lambda_p^2.$$

Eliminating $V_1$ and $V_2$, we have (2.3.11).

## 2.3.4 Extension Beyond Affected Siblings

To extend affected siblings to other affected family members requires a large amount of computation as we can see from the previous section and §2.2. Whittemore and Halpern (1994 a, b) had developed a computational algorithm to handle a reasonable size of pedigree with several affected members. Here we will discuss only the idea of their statistical test, not the computational algorithm.

Fig. 2.3.4 (I) shows a three generation pedigree with 3 affected members $A$, $C$ and $G$ who share both $a$ and $b$ alleles. The main question is: Is the number of alleles shared by the affected individuals large enough to show sign that there is a linkage between the disease gene and the marker? Let $I = (i_{A1} i_{A2}, i_{C1} i_{C2}, i_{G1} i_{G2}) = (ab, ab, bb)$ denote the marker allele configuration of the affected members in the order of $A$, $C$, $G$. The number of common alleles alone in $I$ cannot determine the linkage because the shared alleles is affected by the mating type. For example, the $b$ allele shared by all individuals may not be the same $b$ allele on the same halploid. Thus, we have to use $I$ to construct the IBD sharing. Let $\Phi = (\phi_{A1} \phi_{A2}, \phi_{C1} \phi_{C2}, \phi_{G2} \phi_{G2})$ be the IBD configuration (unknown) of the affected individuals $A$, $C$ and $G$. From IBD viewpoint, there are 6 original marker alleles in this pedigree. They are denoted by (12), (34), and (56) for the founders $A$, $B$, and $E$ respectively in Fig. 2.3.4 (II). Thus, all the values in $\Phi$ should be one of these 6 numbers. Since $\Phi$ is unknown, the best we can do is to

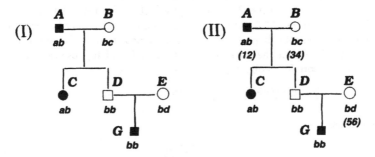

Figure 2.3.4: An example of affected pedigree analysis.

estimate its distribution, i.e. to find $P(\Phi|F)$, where $F$ denote the pedigree information. To compute $P(\Phi|F)$ can be time consuming and storage demanding for a large pedigree, but for a small pedigree such as Fig. 2.3.4, we can find it by inspection.

By the founder's alleles denoted in Fig. 2.3.4 (II), it can be shown that

$$P(\Phi = (12, 13, 25)|F) = P(C = (13)) \cdot P(G = (25)) = 1 \cdot 1/2 = 1/2$$
$$P(\Phi = (12, 13, 35)|F) = P(C = (13)) \cdot P(G = (25)) = 1 \cdot 1/2 = 1/2,$$
(2.3.19)

and all the other configurations of $\Phi$ has probability 0. To summarize the distributed information of $\Phi$ is not easy because they are high-dimensional vectors and the average of the allele numbers has no genetic meaning. We need a score function to measure the IBD content of $\Phi$. Though there can be many choices of score function, we require, at least, the value of a score function increases as more IBDs are shared by the affected individuals. Two such functions $S_1(\Phi)$ and $S_2(\Phi)$ are suggested in Whittemore and Halpern (1994a).

Let $\Phi = (\phi_{11}\phi_{12}, \phi_{21}\phi_{22}, \ldots, \phi_{k1}\phi_{k2})$, in general, contain $k$ individuals. The first function is defined as the average number of pairwise IBD, or

$$S_1(\Phi) = \frac{2}{k(k-1)} \sum_{1 \leq i < j \leq k} f_{ij}(\Phi),$$
(2.3.20)

where

$$f_{ij}(\Phi) = \frac{1}{4} \left[ (\delta(\phi_{i1}, \phi_{j1}) + \delta(\phi_{i1}, \phi_{j2}) + \delta(\phi_{i2}, \phi_{j1}) + \delta(\phi_{i2}, \phi_{j2}) \right]$$

and $\delta(x, y) = 1$ if $x = y$ and $\delta(x, y) = 0$ if $x \neq y$. The second function is defined as

$$S_2(\Phi) = 2^{-k} \sum_u h(u)$$
(2.3.21)

where $u = (u_1, u_2, , u_k)$ with $u_i$ being either $\phi_{i1}$ or $\phi_{i2}$ and $h(u)$ is the number of permutations of the $k$ symbols $(u_1, u_2, \ldots, u_k)$ that leave $u$ unchanged. Suppose $u$ contains $m$ distinct alleles with counts $c_1, c_2, \ldots, c_m (\Sigma c_i = k)$. Then

$$h(u) = \prod_{i=1}^{m} (c_i!) - 1.$$

For example, $h(1, 1, 1) = 3! - 1 = 5$, $h(1, 1, 2) = (2!)(1!) - 1 = 1$, $h(1, 2, 3) = 0$. Both $S_1(\Phi)$ and $S_2(\Phi)$ are intuitive. They are equivalent to the usual IBD score for two affected siblings used in (2.3.3), but neither can dominate the other in terms of power when different alternative hypotheses are used. Suppose that the score function, now denoted by $S(\Phi)$ without the subscript, is determined, and suppose there are $n$ pedigrees. Then the score for pedigree $j$ can be estimated as

$$T_j = E[S(\Phi|F_j)] = \sum_\Phi S(\Phi)P(\Phi|F_j), \quad j = 1, 2, \ldots, n$$
(2.3.22)

and we can use

$$T = \sum_{j=1}^{n} T_j$$
(2.3.23)

as a test statistic and reject $H_0$ (no linkage) if $T$ is much higher than what is expected from $H_0$ (see Exercise 2.13 for other options to summarize the pedigree information). Since $T$ is usually a large sum of independent $T_j$, we can use the central limit theorem to find the $p$-value of rejecting $H_0$ if we can find the mean $E[T_j|F_j, H_0]$ and variance $V[T_j|F_j, H_0]$ of $T_j$ under $H_0$. This can be accomplished by calculating $P(\Phi|F_j, H_0)$. For example, to find $P(\Phi = (12, 13, 25)|F_1, H_0)$ for the configuration in (2.3.19). To do this, we have to **ignore the marker information** on $C$ and $G$, because they might not come under $H_0$. We denote the pedigree without the marker information on the affected individuals as $F_j^\circ$. However, we may either use the marker information $D = (bb)$ or not use it. Whittemore and Halpern prefers the former because it makes the computation much simpler. In this case,

$$P(\Phi = (12, 13, 25|F_1^\circ, H_0) = P(C = (13))P(D = (25)) = 1/4 \times 1/4 = 1/16.$$

A list of $P(\Phi|F_1^\circ, H_0)$ is given in Table 2.3.11. Since the observed value $T_1$ and expected value $E[T_1|F_1^\circ, H_0]$ are similar, the shared alleles among the affected members in this particular pedigree may not support linkage.

Whittemore and Halpern (1994a) did some power comparisons between the two score functions and the mle using (2.2.2) type of likelihood equation for 10 pedigrees each consisting of one affected parent and two affected offspring with full marker information. As expected, the nonparametric test (2.3.23) is less efficient than mle

Table 2.3.11: Computation of $T_1$ by (2.3.22) and of $E[T_1|F_j^\circ, H_0]$ with $P(\Phi|F_j)$ in (2.3.22) being replaced by $P(\Phi|F_j^\circ, H_0)$.

| $\Phi$ | $P(\Phi|F_1, H_0)$ | $P(\Phi|F_1)$ | $S_1(\Phi)$ | $S_2(\Phi)$ |
|---|---|---|---|---|
| (12, 13, 25) | 1/16 | 1/2 | 1/6 | 1/2 |
| (12, 13, 35) | 1/16 | 1/2 | 1/6 | 1/2 |
| (12, 13, 26) | 1/16 | 0 | 1/6 | 1/2 |
| (12, 13, 36) | 1/16 | 0 | 1/6 | 1/2 |
| (12, 14, 25) | 1/16 | 0 | 1/6 | 1/2 |
| (12, 14, 35) | 1/16 | 0 | 1/12 | 1/4 |
| (12, 14, 26) | 1/16 | 0 | 1/6 | 1/2 |
| (12, 14, 36) | 1/16 | 0 | 1/12 | 1/4 |
| (12, 23, 25) | 1/16 | 0 | 1/4 | 1 |
| (12, 23, 35) | 1/16 | 0 | 1/6 | 1/2 |
| (12, 23, 26) | 1/16 | 0 | 1/4 | 1 |
| (12, 23, 36) | 1/16 | 0 | 1/6 | 1/2 |
| (12, 24, 25) | 1/16 | 0 | 1/4 | 1 |
| (12, 24, 35) | 1/16 | 0 | 1/12 | 1/4 |
| (12, 24, 26) | 1/16 | 0 | 1/4 | 1 |
| (12, 24, 36) | 1/16 | 0 | 1/12 | 1/4 |
| $T_1$ (Observed) | | | 0.167 | 0.50 |
| $E[T_1|F_1^\circ, H_0]$ | | | 0.167 | 0.5625 |
| $V[T_1|F_1^\circ, H_0]$ | | | 0.0037 | 0.0791 |

when the likelihood function can be correctly specified, but (2.3.23) becomes more reliable when the penetrance is misspecified. The score function $S_2(\Phi)$ is highly efficient compared to mle for the cases considered. A computer program is available for this analysis (see Sibpair in Appendix B.5).

## 2.4   Designs in Linkage Analysis

### 2.4.1   Design for Simple Mating Types

There is a big difference in linkage analysis design between human genetics and animal or plant breeding. In animal breeding, it is usually possible to design the mating type and compute ELOD or power of the design by standard probability distributions. But for human linkage studies, the mating types cannot be designed. The pedigrees usually have different sizes and structures. Thus, in the design stage, the task is to decide whether to include a particular family or not. Let us start with an animal model. There are many types of matings. Three of them are given in Fig. 2.4.1.

Two parameters are usually used to evaluate the power of a design. One is the expected LOD score discussed in (2.3.2) and the other is the traditional power measured by the type II error in hypothesis testing. Let $z_o$ be the threshold for the LOD score to claim that $\theta$ is less than $1/2$. Then the power is defined as

$$\beta = P\{Z_{\max} \geq z_o\}. \tag{2.4.1}$$

Suppose the mating is the the double backcross in Fig. 2.4.1 with complete penetrance. Then testing

$$H_0 : \ \theta = \theta_o \equiv 1/2 \text{ against } H_1 : \ \theta < 1/2,$$

is the same as testing the binomial probability. As we have seen in §2.2.3, if the significance level is set at $\alpha$ and the power at $1 - \beta$ for the true $\theta = \theta^*$, then the sample size should be

$$n = \frac{\left[z_\alpha \sqrt{\theta_o(1 - \theta_o)} + z_\beta \sqrt{\theta^*(1 - \theta^*)}\right]^2}{(\theta^* - \theta_o)^2}.$$

Since

$$P\{Z_{\max} \geq z_o\} = P\{\chi_1^2 \geq 4.6z_o\}/2$$
$$= P\{N(0, 1) \geq \sqrt{4.6z_o}\},$$

and $\theta_o = 1/2$, we have

$$n = \frac{[1.073\sqrt{z_o} + z_\beta \sqrt{\theta^*(1 - \theta^*)}]^2}{(\theta^* - 0.5)^2}. \tag{2.4.2}$$

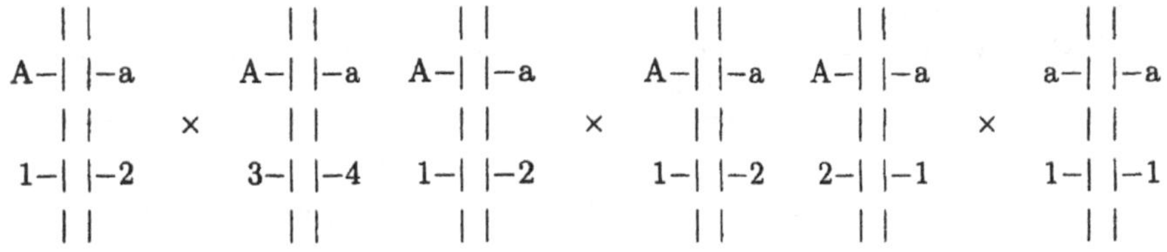

$$
\begin{array}{ccccccc}
|\ | & & |\ | & |\ | & & |\ | & |\ | & & |\ | \\
A-|\ |-a & & A-|\ |-a \quad A-|\ |-a & & A-|\ |-a & A-|\ |-a & & a-|\ |-a \\
|\ | & \times & |\ | \qquad |\ | & \times & |\ | & |\ | & \times & |\ | \\
1-|\ |-2 & & 3-|\ |-4 \quad 1-|\ |-2 & & 1-|\ |-2 & 2-|\ |-1 & & 1-|\ |-1 \\
|\ | & & |\ | \qquad |\ | & & |\ | & |\ | & & |\ |
\end{array}
$$

(1) Full marker information     (2) Double intercross     (3) Double backcross

Figure 2.4.1: Some commonly used mating types. The $a$ and $A$ are alleles of the target gene and $1, 2, 3, 4$ are marker alleles.

**Example 2.4.1** *In a double backcross mating experiment, what is the appropriate sample size to claim linkage between two loci if we set the LOD threshold at 3 and would like to have power 0.8 at $\theta^* = 0.1$?*

**Sol:** In (2.4.2) $z_o = 3$, $\theta^* = 0.1$ and $z_\beta = z_{0.2} = 0.84$. Thus,

$$
n = \frac{[1.073\sqrt{3} + 0.84\sqrt{0.1 \times 0.9}]^2}{(0.1 - 0.5)^2} = 27.8.
$$

A sample with a size of 28 should be enough.    □

For other type of matings, the sample size can be determined similarly. The most informative mating is Type (1) in Fig. 2.4.1. In this arrangement, two crossovers can be observed in every offspring. Thus, it requires only half the sample size of a phase-known double backcross. However, Type (1) mating in Fig. 2.4.1 is not easy to obtain in animal breeding. If we review Mendel's experiment, we see the experiment starts with a pure brand. Pure brand can be obtained by many generations of backcrossing. Unfortunately, when gene A (or a) becomes pure, the markers becomes pure too. Therefore, it is unlikely to have Type (1) breeding in Fig. 2.4.1.

The situation becomes more complicated if the phase is unknown. For example, a phase-unknown double backcross mating has no linkage information if there is only one child. For two children, the probabilities of getting two offspring with various types are given in Table 2.4.1. Note that the total probability $4f_{11} + 4f_{12} + 4f_{21} + 4f_{22} = 1$. There seems no direct connection between the likelihood ratio test and the binomial distribution that could lead to a simple formula like (2.4.2). Since the LOD scores are additive, the distribution of Z in many families follows approximately a normal distribution. In the particular example in Table 2.4.1, E[Z] and Var[Z] at any given $\hat\theta = $ true $\theta^*$ (see the notation in (2.3.2)) can be computed from

$$
E[Z] = \left[\theta^{*2} + (1 - \theta^*)^2\right] \log_{10}\left(\frac{\theta^{*2} + (1 - \theta^*)^2}{0.5}\right) + 2\theta^*(1 - \theta^*) \log_{10}\left(\frac{\theta^*(1 - \theta^*)}{0.25}\right)
$$

Table 2.4.1: Genotypes of two offspring with phase unknown backcross mating (Aa)/(12) × (aa)/(11). The symbols $f_{ij}$ in the parentheses represent the probabilities in front of them.

|  | | Child 2 | |
|---|---|---|---|
|  | | (Aa)/(11) or (aa)/(12) | (Aa)/(12) or (aa)/(11) |
| Child | (Aa)/(11) or (aa)/(12) | $[\theta^2 + (1-\theta)^2]/8$ $(f_{11})$ | $\theta(1-\theta)/4$ $(f_{12})$ |
| 1 | (Aa)/(12) or (aa)/(11) | $\theta(1-\theta)/4$ $(f_{21})$ | $[\theta^2 + (1-\theta)^2]/8$ $(f_{22})$ |

and

$$
E[Z^2] = \left[\theta^{*2} + (1-\theta^*)^2\right] \left[\log_{10}\left(\frac{\theta^{*2} + (1-\theta^*)^2}{0.5}\right)\right]^2
$$
$$
+ 2\theta^*(1-\theta^*)\left[\log_{10}\left(\frac{\theta^*(1-\theta^*)}{0.25}\right)\right]^2
$$
(2.4.3)

For the ideal situation $\theta = \theta^* = 0$, $E[Z] = \log_{10} 2$ and $Var[Z] = 0$. If the required ELOD is set at 3, the number of families needed is

$$
n = 3/\log_{10} 2 \approx 10.
$$

In other words, we need 10 families each with 2 children to reach the required *ELOD*. The small $n$ requirement in this example is because we have luckily selected a marker that is on the gene and get a $Var[Z] = 0$. If we repeat the power requirement of Example 2.4.1 when $\theta^* = 0.1$, we have

$$
E[Z] = 0.096, \quad E[Z^2] = 0.0732, \quad \text{or} \quad \sigma_Z = 0.25. \tag{2.4.4}
$$

The sample size $n$ required should satisfy

$$
Pr\left\{\sum_{i=1}^{n} Z_i \geq 3\right\} = 0.8, \text{ or}
$$
$$
Pr\left\{N(0,1) \geq \frac{3 - 0.096n}{0.25\sqrt{n}}\right\} = 0.8.
$$

Solving a simple quadratic equation, we have $n = 46$. This means 46 families or 92 children, considerably higher than the sample size 28 when the phase is known as in Example 2.4.1.

It can easily be seen that to compute the power analytically will become cumbersome if there are three children or more in the phase-unknown double backcross mating. Simulation becomes one of the most viable methods to estimate the power. The basic steps are:

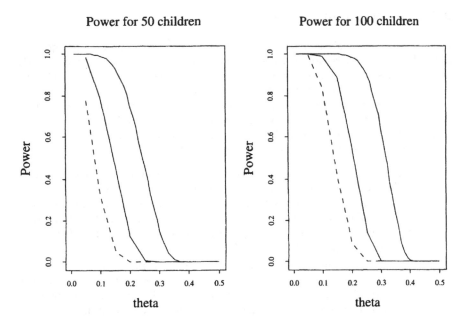

Figure 2.4.2: Power for double backcross families. Top line: phase known (family sizes irrelevant), 2nd from the top: phase-unknown with 4 children per family, 3rd from the top (dotted line): phase-unknown with 3 children per family, left line (dash line): phase-unknown with 2 children per family.

Step 1. Use simulation to estimate the probabilities of each possible offspring combinations with $\theta = 1/2$. Suppose there are $k$ of them. Let them be $p_i(0.5)$, $i = 1, 2, \ldots, k$, e.g., in Table 2.4.1, there are $k = 16$ possible offspring combinations.

Step 2. Again, use simulation to estimate the probabilities of various offspring combinations with a given $\theta$. Let the probabilities be $p_i(\theta)$, $i = 1, 2, \ldots, k$.

Step 3. Compute E[Z] by $\sum_{i=1}^{k} p_i(\theta) \log_{10} \frac{p_i(\theta)}{p_i(0.5)}$.

Step 4. $E[Z^2]$ and Var[Z] can be computed similarly.

The power can then be computed according the central limit theorem as in the previous example. If a computer program for pedigree analysis is available and it can be used repeatedly in simulation, then we need only to simulate the pedigree and find the Z-score for the pedigree. After simulation, the mean and variance of Z can be estimated. Fig 2.4.2 shows the power of various double backcross experiments. When the total number of children is fixed, the power curves confirm that larger families provide more information in the phase unknown case. A computer program to do this is described in Appendix B.2.

## 2.4.2   Selection of Pedigrees

In human linkage analysis, information is usually obtained from large pedigrees of affected families. In this case, we know both the pedigree structure and the individuals who are affected by the disease. Suppose we have a pedigree like that in Fig. 2.4.3. The question now is whether we should use this family in our linkage analysis. Can it provide enough linkage information that it is worth the effort in doing all the marker typing? Usually marker typing is costly, especially when the relatives are scattered in different locations. Boehnke (1986) and Ploughman and Boehnke (1989) worked out the evaluation by a simulation method using Fig. 2.4.3 as an example.

Step 1. The founders' markers are assumed to be independent of the phenotype. This is definitely true if the markers are those junk DNA VNTR. In this case, we may assume that the markers of those founders (1, 2, 8, and 9 in Fig. 2.4.3) satisfy HWE. For example, suppose there are three markers 1, 2, and 3 with prevalence frequencies $f_1$, $f_2$ and $f_3$. Then the chances of founder 1 (or 2, 8, 9) having genotypes 11, 12, etc. are $f_1^2$, $2f_1f_2$, etc.

Step 2. Once the genotype of the founders is determined, the relation between the disease allele and the markers follows the meiotic recombination fraction $\theta$. The offspring genotypes can be simulated accordingly. The first simulated offspring that satisfies the phenotype of that individual will then be accepted. Note in the simulation, the phases are all known (but they may be assumed to be unknown in the data analysis when computing the power.) For example, suppose that in Fig. 2.4.3 the disease gene D is dominant. If founder 1 is $D1/d2$ and 2 is $(d2)/(d3)$ then with probability $(1-\theta)/4$, individual 3 is $(D1)/(d2)$ or $(D1)/(d3)$, and with probability $\theta/4$ she is $(D2)/(d2)$ or $(D2)/(d3)$.

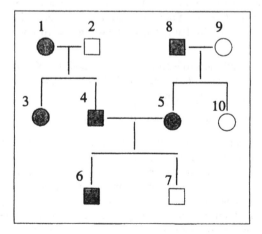

Figure 2.4.3: An example for power simulation.

Step 3. When the pedigree is simulated, compute its LOD score. When this is done many times, we have the mean, the variance and the distribution of the LOD score that may come from this pedigree.

The simulation is simple when the genotype can be uniquely determined by the phenotype. In the example of Fig. 2.4.3, the genotype of founder 2 is definitely $dd$, but the genotype of 1 can be $DD$ or $Dd$. Even when the population frequencies for $D$ and $d$ are known to be $f_D$ and $f_d$, we cannot assume that the probability of founder 1 having $DD$ or $Dd$ satisfy $P(DD) : P(Dd) = f_D^2 : 2f_D f_d = f_D : 2f_d$ because the two offspring are both affected. In this case, it can be shown that $P(DD) : P(Dd) = 2f_D : f_d$. This is still a fortunate case because the genotype of individual 4 has to be $Dd$ from his father or children. If individual 4 could be $DD$ or $Dd$, the distribution of 1's genotype goes to the next level, because we need the distribution of the genotype of individual 4. In general, what we wish to find is the distribution of genotypes $g = \{g_1, g_2, \ldots, g_n\}$ of all the $n$ individuals given their phenotype $x = \{x_1, x_2, \ldots, x_n\}$, or

$$P(g|x). \tag{2.4.5}$$

The following notation is usually used to dissect a pedigree.

$U(i) =$ Upper section of individual $i$, including $i$'s parents and siblings,

$L(i) =$ Lower section of individual $i$, including his offspring, spouse, and the relatives of his/her spouse.

For example, in Fig. 2.4.3, $U(4) = \{1, 2, 3\}$, and $L(3) = \{5, 6, 7, 8, 9, 10\}$. The computation of $P(g|x)$ can be accomplished sequentially as follows:

$$P(g_1, g_2, \ldots, g_n | x_1, x_2, \ldots, x_n) = P(g_1 | x_1, x_2, \ldots, x_n) P(g_2, g_3, \ldots, g_n | g_1, x_1, x_2, \ldots, x_n)$$
$$= P(g_1 | x_1, x_2, \ldots, x_n) P(g_2 | g_1, x_1, x_2, \ldots, x_n)$$
$$P(g_3, g_4, \ldots, g_n | g_1, g_2, x_1, x_2, \ldots, x_n)$$
$$= \cdots$$

Since for a given genotype the phenotype plays no role in inheritance, the general term can be simplified by

$$P(g_k | g_1, g_2, \ldots, g_{k-1}, x_1, x_2, \ldots, x_n) = P(g_k | g_1, g_2, \ldots, g_{k-1}, x_k, x_{k+1}, \ldots, x_n).$$

If the array $g_1, g_2, \ldots, g_{k-1}$ contains the parents of $g_k$, say that they are $g_i$, $g_j$, then

$$P(g_k | g_1, g_2, \ldots, g_{k-1}, x_{k+1}, x_{k+2}, \ldots, x_n)$$
$$= P(g_k | g_i, g_j, x_k, x_{L(k)})$$
$$= \frac{P(x_k | g_k) P(x_{L(k)} | g_k) P(g_k | g_i, g_j)}{\sum_{\substack{All\ possible\ g_k^*}} P(x_k | g_k^*) P(x_{L(k)} | g_k^*) P(g_k^* | g_i, g_j)}. \tag{2.4.6}$$

All the probabilities can be computed easily except $P(x_{L(k)} | g_k)$. Let us use $P(x_1, x_2, \ldots, x_n | g_1)$ as an example. If we go back to (2.2.2), we see that $P(x_1, x_2, \ldots, x_n | g_1)$ is the same as the right side of (2.2.2) except that $g_1$ is given and the summation over $g_1$ is unnecessary. Since we know how to do (2.2.2), we should have no problem computing $P(x_1, x_2, \ldots, x_n | g_1)$, although one has to write a computer program to do it.

Table 2.4.2: Power for the two pedigrees in Fig. 2.4.4.

*Mean maximum LOD score ± standard error for a dominant disease*
*and a linked codominant marker*

| True recombination fraction $\theta$ | Pedigree | | |
|---|---|---|---|
| | A | B | A, B combined |
| .00 | $2.84 \pm .04$ | $2.57 \pm .04$ | $5.41 \pm .06$ |
| .05 | $2.07 \pm .04$ | $1.92 \pm .04$ | $3.84 \pm .06$ |
| .10 | $1.52 \pm .04$ | $1.45 \pm .03$ | $2.75 \pm .05$ |
| .15 | $1.05 \pm .03$ | $1.02 \pm .03$ | $1.89 \pm .04$ |
| .20 | $0.70 \pm .02$ | $0.77 \pm .02$ | $1.28 \pm .03$ |

*Note*: Means and standard errors are based for each $\theta$ on $N = 1000$ simulated pedigrees assuming a two-allele codominant marker with allele frequencies 0.50. (From Boehnke (1986), copyright ©by The University of Chicago Press and the author. Reprinted by permission.)

Not all the conditional probabilities are as simple as (2.4.6), because $k$'s parents may not be contained in $g_1, g_2, \ldots, g_{k-1}$. Several cases that have to be considered differently are given in Ploughman and Boehnke (1989). Fortunately, in many practical cases, the disease gene is a rare gene. Thus the genotype is *almost* determined by the phenotype. If we neglect the small uncertainty of the determination of the genotype, then we can return to the simple simulation case by ignoring the complicated computation (2.4.5–2.4.6). With this approximation in mind, if the disease gene is rare dominant ($D$), then an affected person is $Dd$ and an unaffected person is $dd$. On the other hand, if the disease gene $d$ is rare recessive, then an affected person is $dd$, and an unaffected is $Dd$ if one of his/her offspring has the disease.

**Example 2.4.2**   *(Boehnke, 1986). Estimate the power of linkage detection by LOD method on the following two pedigrees, assuming the disease is caused by a rare dominant gene and the marker is codominant with equal allele frequency 0.5.*

**Sol**: The simulated power is given in Table 2.4.2. A simulation program can be found in Appendix B.4.                                                                      □

## 2.4.3   Design in Sibpair Study

The power for sibpair tests can be obtained more easily due to the simple pedigree structure. When (2.3.5) is used, the power is

$$1 - \Phi\left(\frac{z_\alpha - \sqrt{2n}(p_2 - p_0)}{\sqrt{2(p_0 q_0 + p_2 q_2 + 2p_0 p_2)}}\right),$$

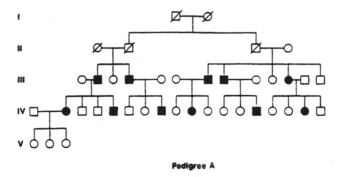

**Pedigree A**

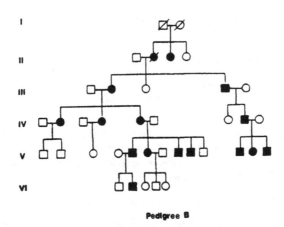

**Pedigree B**

Figure 2.4.4: Two pedigrees for Example 2.2.4 (From Boehnke(1986), copyright ©by The University of Chicago Press and the author. Reprinted by permission.)

where $\alpha$ is the significance level, $\Phi()$ is the standard normal distribution function, $p_i = Pr\{S_m = i\}$ with $S_m$ defined in (2.3.3), $q_i = 1 - p_i$, for $i = 0$ and 2. The derivation is based on the asymptotic distribution of a multinomial distribution (see Appendix A.3). The computation of the $p_i$'s depends on the recombinational fraction $\theta$ and the penetrance. They can be found in (2.3.16). Some details are left as an exercise.

The power of the N test defined by (2.3.10) has been worked out by Green and Shah (1992). The formula depends on six parameters: the total sib size, number of affected sibs, the penetrance for a person having 0, 1 or 2 disease alleles, and the disease allele frequency in the general population. Though the analytic formula is complicated, the power can obtained numerically by simulation.

To design the sibpair study based on (2.3.6) and (2.3.12) is also straightforward, although it can be computationally intensive. Referring back to the notation in this test, for given $\lambda_p$, $\lambda_o$, $\lambda_s$ and $\theta$, the three $p_i$'s are fixed. Thus, the distribution of $n_0$, $n_1$

and $n_2$, conditional on a total sample size $n_0 + n_1 + n_2 = n$, follows a multinomial distribution (see Appendix A.3), i.e.,

$$P(n_0, n_1, n_2) = \frac{n!}{n_0! n_1! n_2!} p_0^{n_0} p_1^{n_1} p_2^{n_2}. \qquad (2.4.7)$$

The LOD score from the likelihood ratio test has the value

$$Z(\hat{\theta}) = \max_\theta \sum_{i=0}^{2} n_i \log_{10} p_i(\theta) - \sum_{i=0}^{2} n_i \log_{10} p_i(\theta = 0.5). \qquad (2.4.8)$$

If $n$ is not large, the distribution of LOD can be computed and so can ELOD and the power $\Pr\{Z(\hat{\theta}) \geq z_o\}$, where $z_o$ is the threshold in the LOD test. For large $n$, normal approximation can be used as in the following:

$$Pr\{I - 1/2 \leq n_o \leq I + 1/2 \quad \text{and} \quad J - 1/2 \leq n_1 \leq J + 1/2\} \approx \int_{I-1/2}^{I+1/2} \int_{J-1/2}^{J+1/2} f(x, y) dx dy,$$

where $f(x, y)$ is the density function of bivariate normal

$$N\left( \begin{pmatrix} np_0 \\ np_1 \end{pmatrix}, \begin{pmatrix} np_0 q_0 & -np_0 p_1 \\ -np_0 p_1 & np_1 q_1 \end{pmatrix} \right),$$

where $q_i = 1 - p_i$.

Risch (1990b) computed the power when the likelihood function was based on $n_0$ (or $n - n_0$) only. In this case, the multinomial (2.4.7) becomes binomial and the bivariate normal approximation becomes the univariate normal approximation. It can be shown that $Z(\hat{\theta})$ in (2.4.8) becomes

$$Z(\hat{\theta}) = n_0 \log_{10}(4n_0/n) + (n - n_0) \log_{10}[4(n - n_0)/(3n)].$$

Fig. 2.4.5 shows the power for $\theta = 0$ with an extremely polymorphic marker. Note that the $\lambda_s$ in these two figures are actually $\lambda_s/\lambda_p$ in our notation.

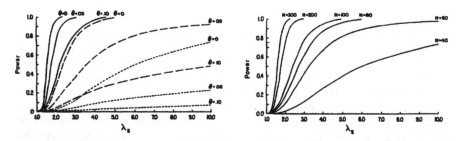

Figure 2.4.5: Power to detect linkage as a function of $\lambda_s$, affected sibpair size $N$, and recombination fraction $\theta$. The marker is assumed to be fully informative. In the left figure, solid lines, $N = 300$; dashed lines, $N = 100$; dotted lines, $N = 40$. In the right figure, $\theta = 0$. (From Risch (1990b), copyright ©by the University of Chicago Press. Reprinted by permission.)

### 2.4.4 Other Concerns in Pedigree Analysis and Design

There are many complications in large pedigree analysis. It is not the intention of this book to cover all of them, but a few interesting ones are mentioned in this session. A more comprehensive treatise in this area can be found in Ott (1991).

One difficult situation in large pedigree analysis is marriage loops, i.e., marriages between relatives. Fig. 1.1.2 is redrawn in Fig. 2.4.6 with a few hypothetical marriage loops added (in dotted lines). Originally, when conditioned on the genotypes of individuals $III(1), III(4)$, and $III(26)$, the whole pedigree can be decomposed into three independent small pedigrees. However, this decomposition is no longer valid when loops exist. How to deal with loops can be found in Thompson (1986) and Terwilliger and Ott (1994).

We have mentioned sex-linked inheritance in 2.1.3. There is evidence that the male and female recombination rates may not be the same between two particular loci (Renwick and Schulz (1965)). Thus, in Table 2.2.1 two combination fractions $\theta_m$ and $\theta_f$ should be used for the mother and the father respectively. Besides that one more parameter is added to the likelihood equation, there is no theoretical complication in findings the mle of the two-parameters, especially when the maximization is done by exhaustive search. The two generation linkage analysis computer program in Appendix B.1 can handle this situation.

As we have noticed in the diabetes example (see comments on Example 2.2.1), the disease gene may vary among families. The reason is quite simple. To make a physiological function normal, many enzymes may be necessary. Any gene that controls these enzymes can be considered as a disease gene. Hence for a particular marker, the recombination fraction $\theta$ with different genes can be different. Morton (1956) used the likelihood ratio principle to test the hypotheses $H_0 : \theta < 1/2$ and $\theta$ is uniform among the families against $H_1 : \theta$'s are heterogeneous among the families. This alternative

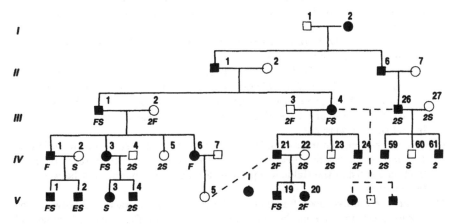

Figure 2.4.6: Example of loops (in dotted lines) in a pedigree.

was later made to be more specific by Smith (1961, 1963) with the assumption that there are only two or three different $\theta$'s.

In linkage analysis, pedigrees may not be collected all at once. They can be collected sequentially. Morton (1955) applied Wald's (1947) sequential method in linkage analysis, and found that it was more efficient than fixed sample methods when double backcross sibpairs could be obtained sequentially. The fixed sample methods he compared were Fisher-Finney's $u$ test (Fisher, 1935 and Finney, 1940) and Haldane and Smith (1947) likelihood ratio test. However, when linkage analysis is not based on sibpairs, additional family members are often gathered only when we think those members can provide substantial linkage information. For this type of subjective decision, classic sequential method is difficult to apply. We may ignore the sequential decision part and ask only one question: when data are gathered by sequential decisions, is the final result based on one likelihood equation using all the data biased? Cannings and Thompson (1977) showed that the inference on the mode of inheritance of a characteristic is not biased provided that:

(i)  the choice of individuals to be examined next depends only on the phenotypes already observed, and

(ii) all individuals (pedigrees) who are examined are include in the analysis.

These two requirements are not too difficult to meet.

# Exercise 2

2.1 How many phase-known meioses do we need to confirm $P(\theta \leq 0.01) \approx 0.95$ in human linkage analysis from a Bayesian viewpoint if all the meioses had no recombination?

2.2 How do you modify the prior distribution used in (2.1.3) if the marker is known on chromosome 22? (Hint: use Table 1.2.2.)

2.3 Construct a table similar to Table 2.1.1 and $L(\theta)$ with the pedigree in Fig.1.4.4, assuming (1) the disease gene $d$ is recessive, (2) the parents and grandparents are all healthy and the phase of the father is known to be $D2/d4$ and the mother is $D1/d3$, and (3) the first three children from the left were affected by the disease, but all the rest were not.

2.4 Two markers $m_1$ and $m_2$ are randomly picked from a chromsome which has length $L$. Suppose the relation between physical distance $x$ and their recombination fraction $\theta$ satisfies the Haldane equation. Find the distribution of $\theta$ between the two markers.

2.5 Determine the likelihood function for the recombination fraction $\theta$ between the disease gene and the marker in pedigrees shown in Fig. E2.5. (Assume full penetrance.)

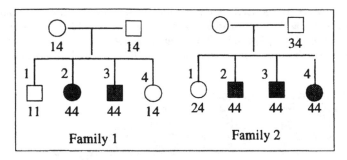

Figure E2.5.

2.6  Let $D$ be a dominant disease allele and $d$ be a healthy one. The mating type of the the parents is the phase-known double backcross $d1/D2 \times d1/d1$. The age affect of the disease onset has been estimated in the following table:

| Genotype | Age (first row), onset rate (2nd and 3rd rows) | | | |
|---|---|---|---|---|
|  | 0−10 | 11−20 | 21−30 | 31+ |
| $D/D$ or $D/d$ | 0.2 | 0.4 | 0.6 | 0.9 |
| $d/d$ | 0.0 | 0.1 | 0.1 | 0.2 |

Find the maximum LOD score of the linkage between the marker (with allele symbols 1 and 2) and the disease gene with the following data gathered from several families. (You will need to write a simple computer program to calculate $Z_{max}$.)

| Age | 4 | 8 | 12 | 15 | 17 | 16 | 20 | 24 | 25 | 35 | 36 | 38 | 40 |
|---|---|---|---|---|---|---|---|---|---|---|---|---|---|
| Affected?* | Y | N | Y | Y | N | N | N | Y | N | Y | Y | N | Y |
| Marker | 11 | 12 | 12 | 12 | 11 | 11 | 11 | 12 | 11 | 12 | 12 | 11 | 11 |

* Y = yes, N = no

2.7  Show that the probability for $ab \times ac$ in Table 2.3.7 is correct.

2.8  Show that the probability $P(\pi_t = 1|\pi_m = 1) = \Psi^2 + (1 - \Psi)^2$ in Table 2.3.10 is correct.

2.9  Show that the probability $p_1$ in (2.3.16) is correct.

2.10 Use *liped* (see Appendix B.3) to find the maximum likelihood estimate of $\theta$ and LOD for the *dentinogensis imperfecta* gene and the GC protein gene from the pedigree in Fig 1.1.2. You may eliminate the first two generations in your analysis.

2.11 The pedigrees in Fig. E2.11 are from families with cystic fibrosis (an autosomal recessive disease) in a research report (Spence et al. (1986)). Use the linkage software *liped* (see Appendix B.3) to analyze the families with five children or more using the first marker with alleles $A$, $B$, $C$, $D$ and $E$. (Ignore the NOP marker).

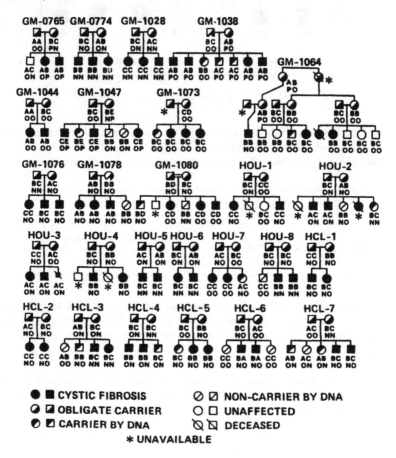

Figure E2.11: Twenty six families affected by cystic fibrosis. (From Spence et al. (1986), copyright ©by The University of Chicago Press. Reprinted by permission.)

2.12 The following three pedigrees (Fig. E2.12) are obtained from patients with a dominant autosomal genetic disease. Suppose the penetrance is 100%. Use *Simlink* (see Appendix B.4) to valuation the potential of using these pedigrees to confirm linkage between the disease gene and a marker which has three alleles with population frequencies 0.5, 0.3 and 0.2. In particular, find the mean and standard deviation of the possible LOD scores for true $\theta$ being 0.0, 0.025, 0.05, 0.10 or 0.50, if all the living individuals are typed.

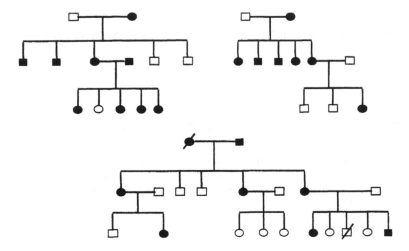

Figure E2.12: Pedigrees for Exercise 12.

2.13 The following three pedigrees (Fig. E2.13) are obtained from patients with a genetic disease of unknown penetrance and inheritance mode. Evaluate the linkage between the disease gene and the marker by Whittemore and Halpern's

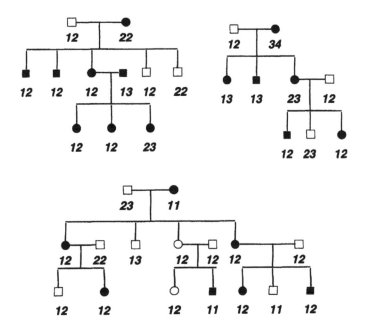

Figure E2.13: Pedigrees for Exercise 13.

nonparametric method using *SIBPAIR* (see Appendix B.5). Note that *SIBPAIR* does not use (2.3.23) to summarize the pedigree information. It uses a weighted sum

$$Z = \frac{\sum_{j=1}^{n} w_j (T_j - E[T_j]) / \sqrt{V[T_j]}}{\sqrt{\sum_{j=1}^{n} w_j^2}},$$

where $E[T_j]$ and $V[T_j]$ are the mean and variance of $T_j$ under $H_0$ and $w_j = \sqrt{k_j - 1}$, where $k_j$ is the number of affected individuals in pedigree $j$. Confirm this formula from the *SIBPAIR* outputs. What should we do when $V[T_j] = 0$?

# Chapter 3

# Genetics of Quantitative Trait

## 3.1 Quantitative Trait Locus (QTL) Based on Sibpairs

Many genetics books, especially in the field of biology or molecular genetics, tend to trace the origin of modern genetics to Mendel's experiments in the 1860s. However, at almost the same time, F. Galton started his work on quantitative genetics, mainly using the stature of human beings (Galton, 1889). Apparently, he noticed that height was inheritable, but the pattern on how stature is inherited is much less clear-cut than Mendel's sweet pea characteristics. Galton, Karl Pearson and their students developed their theory by using correlation and linear model such as

$$X_{n+1} = h \frac{X_n' + X_n''}{2} + \varepsilon_{n+1},$$

where $h$ measures the heritability, $X_{n+1}$ is the trait value in the $(n + 1)$st generation, $X_n'$ and $X_n''$ are the parental trait values in the $n$th generation, and $\varepsilon_{n+1}$ denotes the random variable due to environment (see Karlin, 1992). Although not as precise as Mendel's gene theory, a new branch of science, called biometry, was born.

As soon as Mendel's theory was rediscovered in 1900, biometricians began to use Mendel's model for their data. It turned out that Mendel's one gene one characteristic model did not fit many continuous data. As early as 1909, two Scandinavian geneticists, Johannsen and Nilsson-Ehle discovered that certain continuous traits in wheat and beans could not be controlled by a single gene (Mather and Jinks, 1971). Later Warren (1924), Mather and his colleagues made many similar observations in *Drosophila melanogaster* (Mather, 1941, 1942 and Breese and Mather, 1957, 1960)). The purpose of this chapter is to survey what has been done in linking a quantitative trait to genes. We start with a single gene.

Table 3.1.1: Phenotype distributions by genotypes in QTL.

| Genotype ($g$) | $bb$ | $bB$ | $BB$ | |
|---|---|---|---|---|
| Prevalence | $f_{bb}$ | $f_{bB}$ | $f_{BB}$ | $(f_{bb} + f_{bB} + f_{BB} = 1)$ |
| Phenotype | $\mu_{bb} + \varepsilon_{bb}$ | $\mu_{bB} + \varepsilon_{bB}$ | $\mu_{BB} + \varepsilon_{BB}$ | |

In principal, the general formula (2.2.2) can be used to locate position for a quantitative trait locus (QTL). However, because of the quick loss of linkage information over generations, using sibpair is one of the most commonly used methods in searching for a QTL in human genetics. The situation can be quite different in animal or crop breeding because more genetic information may be available from controlled mating and from a large number of offspring. However, sibpairs serve as a simple introduction to QTL analysis.

In the beginning let there be only two alleles, $B$ and $b$, in a locus. Their prevalence and phenotypes are given in Table 3.1.1, where the $\mu$s are the mean values of the phenotypes and the $\varepsilon$s are random variations with 0 mean. Note that the phenotype variation within the same genotype should be large enough to cause considerable overlapping. Otherwise genotypes can be easily determined by the phenotypes and the data are equivalent to those with a qualitative trait. It is these large phenotype overlaps in different genotypes that make the genetic information difficult to trace over generations.

In the conventional statistical notation, let $X|g$ be the trait values of the genotype $g$, i.e.,

$$X|g = \mu_g + \varepsilon_g, \quad g = bb, bB, BB.$$

In genetics, like in one-way analysis of variance (see §A.2), we rewrite this model as,

$$X|g = \mu + \gamma_g + \varepsilon_g, \quad g = bb, bB, BB, \tag{3.1.1}$$

where $\mu$ is the overall mean and

$$\gamma_g = \begin{cases} -a & \text{for} \quad g = bb \\ d & \text{for} \quad g = bB \\ +a & \text{for} \quad g = BB, \end{cases} \tag{3.1.2}$$

for $a \geq 0$. When $d = 0$ in (3.1.2), the trait is called **additive** because the $\mu_i$ is proportional to the number of $b$ (or $B$) alleles in the locus. When $d \neq 0$ the alleles are said to have a **dominance effect**. For example, $d > 0$ implies that $B$ dominates $b$ and $d < 0$ implies that $b$ dominates $B$. The familiar cases are $d = a$ or $d = -a$, when $B$ or $b$ has total dominance.

Suppose now that a marker is linked to the gene we wish to locate. In this instance, we should expect that the two sibs who share the same marker tend to have similar trait values. Let $X$ and $Y$ denote the trait values of two sibs with $S_m$ being the number of common alleles between them; $S_m$ can be 0, 1 or 2. Thus, if the marker is linked to the trait gene, $|X - Y|$ and $S_m$ are expected to be negatively corrected, and a negative correlation shows linkage. The correlation test (see Appendix A.1),

$$r \sim N(0, 1/(n-3))$$

will provide statistical confidence in the decision, where $r$ is the sample correlation coefficient between $|X - Y|$ and $S_m$ and $n$ is the number of sibpairs in the study. To have more insight into the relation between $|X - Y|$ and $S_m$, we follow the derivation of Haseman and Elston (1972).

Since the absolute value is not as easy to handle as its square, Haseman and Elston define

$$\Delta_j^2 = |X_j - Y_j|^2 \tag{3.1.3}$$

where $j$ is the family index. They also assume the frequencies in Table 3.1.1 have reached HWE, i.e., $f_{bb} = q^2$, $f_{bB} = 2pq$ and $f_{BB} = p^2$ with $p + q = 1$.

Before deriving the relation between $\Delta_j^2$ and the recombination fraction $\theta$, we first introduce two often used variance components in genetics. If we let $X$ denote a random sample from the population under HWE, then

$$E[X] = q^2(\mu - a) + 2pq(\mu + d) + p^2(\mu + a)$$
$$= \mu + ap^2 + 2pqd - aq^2.$$

By assuming that the all the error terms $\varepsilon_g$ in (3.1.1) have the same variance $\sigma_\varepsilon^2$, we have

$$\mathrm{Var}[X] = a^2p^2 + 2pqd^2 + a^2q^2 - (ap^2 + 2pqd - aq^2)^2 + \sigma_\varepsilon^2. \tag{3.1.4}$$

After algebric derivation, the genetic variance component $\sigma_g^2 \equiv \mathrm{Var}[X] - \sigma_\varepsilon^2$ can be simplified to

$$\sigma_g^2 = 2pq[a - d(p - q)]^2 + 4p^2q^2d^2$$
$$\equiv \sigma_a^2 + \sigma_d^2, \tag{3.1.5}$$

where $\sigma_a^2$ and $\sigma_d^2$ are, respectively, the first and second term of the first line in (3.1.5). They are called the **the additive variance** ($\sigma_a^2$) and the **dominant variance** ($\sigma_d^2$). Note that if $d = 0$, then $\sigma_d^2 = 0$ and $\sigma_g^2 = 2pqa^2$. Why $\sigma_g^2$ is partitioned as (3.1.5) will be explained later in the derivation of (3.3.9). The proportion of variation that can be explained by inheritance is called the **heritability** $h^2$, defined as

$$h^2 = \sigma_g^2/(\sigma_g^2 + \sigma_\varepsilon^2) = 1 - \sigma_\varepsilon^2/(\sigma_g^2 + \sigma_\varepsilon^2). \tag{3.1.6}$$

It can be seen that this $h^2$ is similar to the $R^2$ in regression analysis, that is the percentage of variance that can be explained by the genetic information. In the genetic literature, heritability is not always defined the same way. For example, it has also been defined as $h^2 = \sigma_a^2/(\sigma_g^2 + \sigma_\varepsilon^2)$, i.e., only the additive effect is considered inheritable. In general, if $X$ denotes the trait value of an individual, $\sigma_X^2$ denotes the variance of $X$ without any genetic information, and $\sigma_e^2$ denotes the variance of $X$ when the genetic information of the individual is given, then the heritability of this trait can be defined as

$$h^2 = 1 - \sigma_e^2/\sigma_X^2. \tag{3.1.7}$$

The derivation of the relation between $\Delta_j^2$ and $\theta$ follows a formula similar to (2.3.14). As defined in §2.3,

$$
\begin{aligned}
g_t &= \text{genotype of the two sibs (a vector of two values)}\\
\pi_t &= \text{IBD at the trait locus shared by the two sibs}\\
\pi_m &= \text{IBD at the marker locus}\\
S_m &= \text{number of markers shared by the two sibs.}
\end{aligned}
\qquad (3.1.8)
$$

Only $S_m$ is observable. By Bayes' theorem and the fact that $E(\Delta^2|g_t, \pi_t, \pi_m, S_m) = E(\Delta^2|g_t)$, we have

$$
E(\Delta^2|S_m) = \sum_{g_t}\sum_{\pi_t}\sum_{\pi_m} E(\Delta^2|g_t)P(g_t|\pi_t)P(\pi_t|\pi_m)P(\pi_m|S_m). \qquad (3.1.9)
$$

Let the markers be extremely polymorphic, so that $\pi_m = S_m$. Consequently, the last term $P(\pi_m|S_m)$ vanishes with the third summation. The third term $P(\pi_t|\pi_m)$ has already been derived in Table 2.3.10. Thus, we need only to compute the first two terms in (3.1.9). They are given in Table 3.1.2, where $\sigma^2 = E[\varepsilon_X - \varepsilon_Y]^2$ and $\varepsilon_X$ and $\varepsilon_Y$ are the error terms of the two sibs using model (3.1.1). Since $\epsilon_X$ and $\epsilon_Y$ tend to be positively correlated because two sibs were usually raised in the same environment, $\sigma^2$ may not be $2\sigma_\varepsilon^2$. Actually, it can even be less than $\sigma_\varepsilon^2$.

The derivation of the probabilities are easier for $\pi_t = 0$ or 2. Here, only the derivations of two examples when $\pi_t = 1$ are presented.

$$
\begin{aligned}
P(Bb, BB|\pi_t = 1) &= P\{\text{IBD is on the father's side and the non-IBD from}\\
&\qquad \text{the mother's side}\}\cdot\\
&\qquad \cdot P\{\text{IBD is B}\}\cdot P\{\text{ the non-IBDs are } B \text{ and } b\}\\
&\quad + \text{ a symmetric term for the mother's side}\\
&= \frac{1}{2}\cdot p\cdot 2pq + \frac{1}{2}\cdot p\cdot 2pq\\
&= 2p^2 q.
\end{aligned}
$$

Table 3.1.2: Conditional probabilities for the terms in (3.1.9).

| $g_t$ | $E[\Delta^2|g_t]$ | $P(g_t|\pi_t = 0)$ | $P(g_t|\pi_t = 1)$ | $P(g_t|\pi_t = 2)$ |
|---|---|---|---|---|
| BB, BB | $\sigma^2$ | $p^4$ | $p^3$ | $p^2$ |
| bb, bb | $\sigma^2$ | $q^4$ | $q^3$ | $q^2$ |
| Bb, Bb | $\sigma^2$ | $4p^2 q^2$ | $pq$ | $2pq$ |
| Bb, BB | $(a-d)^2 + \sigma^2$ | $4p^3 q$ | $2p^2 q$ | 0 |
| Bb, bb | $(a+d)^2 + \sigma^2$ | $4pq^3$ | $2pq^2$ | 0 |
| bb, BB | $4a^2 + \sigma^2$ | $2p^2 q^2$ | 0 | 0 |
| Total |  | 1 | 1 | 1 |

$P(Bb, Bb|\pi_t = 1) = P\{$IBD is on the father's side and the non-IBD from

the mother's side$\} \cdot P\{$IBD is $B\} \cdot$

$\cdot P\{$the non-IBDs are $b$ and $b\}$

$+ P\{$IBD is on the father's side and the non-IBD from

the mother's side$\} \cdot P\{$IBD is $b\} \cdot$

$\cdot P\{$the non-IBDs are $B$ and $B\}$

$+$ symmetric terms for the IBD on the mother's side

$$= \frac{1}{2}(p \cdot q^2 + q \cdot p^2) + \frac{1}{2}(p \cdot q^2 + q \cdot p^2)$$

$$= pq.$$

In the above derivation, all the probabilities are conditioned on $\pi_t = 1$. The first two terms of (3.1.9) can be combined as

$$E(\Delta^2|\pi_t = 2) = \sum_{g_t} E(\Delta^2|g_t)P(g_t|\pi_t = 2) = \sigma^2,$$

$$E(\Delta^2|\pi_t = 1) = \sigma^2 + (a - d)^2(2p^2q) + (a + d)^2(2pq^2) \qquad (3.1.10)$$

$$= \sigma^2 + \sigma_a^2 + 2\sigma_d^2,$$

$$E(\Delta^2|\pi_t = 0) = \sigma^2 + 2\sigma_a^2 + 2\sigma_d^2.$$

The last equality in (3.1.10) can either be derived from the summation of the conditional probability or by the definition that $X$ and $Y$ are independent. Now (3.1.9) is reduced to

$$E(\Delta^2|S_m) = \sum_{\pi_t} E[\Delta^2|\pi_t]P(\pi_t|\pi_m = S_m). \qquad (3.1.11)$$

Using the $P(\pi_t|\pi_m)$ values in Table 2.3.10, we have

$$m_0 \equiv E(\Delta^2|\pi_m = 0) = \sigma^2(1 - \Psi)^2 + 2\Psi(1 - \Psi)(\sigma^2 + \sigma_a^2 + 2\sigma_d^2) + \Psi^2(\sigma^2 + 2\sigma_a^2 + 2\sigma_d^2)$$

$$= \sigma^2 + 2\Psi\sigma_a^2 + 2\Psi(2 - \Psi)\sigma_d^2,$$

$$m_1 \equiv E(\Delta^2|\pi_m = 1) = \sigma^2 + \sigma_a^2 + 2(\Psi^2 - \Psi + 1)\sigma_d^2,$$

$$m_2 \equiv E(\Delta^2|\pi_m = 2) = \sigma^2 + 2(1 - \Psi)\sigma_a^2 + 2(1 - \Psi^2)\sigma_d^2. \qquad (3.1.12)$$

Recall that $\Psi = \theta^2 + (1 - \theta)^2$. Haseman and Elston (1972) and Elston (1988) use (3.1.12) with the assumption that $d = 0$ to test the hypothesis $\theta = 1/2$. When $d = 0$, (3.1.12) is simplified to

$$m_i = \sigma^2 + 2\Psi\sigma_a^2 + (1 - 2\Psi)\sigma_a^2 \cdot i = 2(1 - 2\theta(1 - \theta))\sigma_a^2 + \sigma^2 - (1 - 2\theta)^2\sigma_a^2 \cdot i.$$

Hence data $(\Delta^2, \pi_m = S_m)$ can be expression by the regression equation

$$\Delta^2 = \alpha + \beta\pi_m + \varepsilon, \qquad (3.1.13)$$

and to test that $\theta = 1/2$ is equivalent to test $\beta = 0$. Since $(1 - 2\Psi)\sigma_a^2$ is the expected slope of the regression equation, the test should be one-sided with $H_0 : \beta = 0$ against

$H_1 : \beta < 0$. Note that the $\Delta^2$ in (3.1.13) is not normally distributed and to use the usually t-test to test the slope is only an approximation. More discussion on using expectation discrepancy as random error is given in §3.2.3.

For not so polymorphic markers, the $\pi_m$ in (3.1.13) is unknown and has to be estimated from the marker information, because the number of identical markers $S_m$ is not sufficient to predict $\pi_m$. Let the full marker information of the two sibs be $I_m$. Obviously, we should estimate $\pi_m$ by

$$\hat{\pi}_m = E(\pi_m|I_m) = 2P(\pi_m = 2|I_m) + P(\pi_m = 1|I_m),   \tag{3.1.14}$$

and replace (3.1.13) with

$$\Delta^2 = \alpha + \beta\hat{\pi}_m + \varepsilon.   \tag{3.1.15}$$

To find $P(\pi_m = i|I_m)$ is not difficult when the parental markers are known. A table is given in Haseman and Elston (1972, Table II). We will illustrate the procedure with a particular example. Suppose the sibs markers are $(ab)$ and $(ab)$ and the parent's types are $aa \times ab$. Using Table 2.3.2 with $1 = 2 = 3 = a$ and $4 = b$, we have $P[I_m = (ab, ab)] = 4/16$, $P[I_m \cap (\pi_m = 1)] = P[I_m \cap (\pi_m = 2)] = 2/16$. Thus, $P(\pi_m = 2|I_m) = P(\pi_m = 1|I_m) = 1/2$, or $\hat{\pi}_m = 3/2$. When parental markers are unknown, $\hat{\pi}_m$ can be computed with the help of (2.3.8).

**Example 3.1.1** *Table 3.1.3 gives five sibpairs with their marker types and quantitative values Y. Can we claim linkage between the marker and the QTL?*

Table 3.1.3: Marker genotypes and quantitative values of five sibpairs.

| Family | Parental mating type | Marker & Y for sib 1 | Marker & Y for sib 2 |
|--------|---------------------|---------------------|---------------------|
| 1 | $12 \times 13$ | 12, 55 | 12, 58 |
| 2 | $11 \times 34$ | 13, 54 | 14, 64 |
| 3 | $12 \times 34$ | 13, 65 | 14, 60 |
| 4 | $11 \times 12$ | 12, 57 | 12, 51 |
| 5 | $12 \times 23$ | 12, 58 | 23, 50 |

**Sol**: From this table, it is not difficult to estimate $\pi_m$. The results are given as follows.

| Family | 1 | 2 | 3 | 4 | 5 |
|--------|---|---|---|---|---|
| $S_m$ | 2 | 1 | 1 | 2 | 1 |
| $\hat{\pi}_m$ | 2 | 0.5 | 1 | 1.5 | 0 |
| $\Delta^2$ | 9 | 100 | 25 | 36 | 64 |

The regression equation (3.1.13) becomes

$$\Delta^2 = 81.6 - 34.8\pi_m + \varepsilon,$$

where the standard error of $\hat{\beta}$ is 16.809, which gives a T-test value $-2.07$ and one sided $p$-value $0.0651$.

A computer program to do data analysis based on (3.1.13) is given in Appendix B.5.

Though (3.1.13) and (3.1.14) lay a foundation for human QTL search, they still require a very large sample size to reach reasonable accuracy. Blackwelder and Elston (1982) did some power analysis using a formula similar to (3.1.13) to test the hypothesis $\theta = 1/2$. Instead of highly polymorphic markers, the authors assumed that the marker had only two alleles with prevalence frequency $p = 0.5$. Fig. 3.1.1 gives some idea of the sample size requirement under the very moderate significance level $= 0.05$ and power $= 0.90$. These curves were obtained using asymptotic theory and checked by simulation.

In Fig. 3.1.1, heritability $h^2$ is defined as $\sigma_g^2/(\sigma_g^2 + \sigma^2/2)$. Since $\sigma_\epsilon^2$ and $\sigma^2 = E[\epsilon_X - \epsilon_Y]^2$ are not simply related, the authors assumed $\sigma^2 = \sigma_\epsilon^2/2$. Since $\sigma_g^2 = 2pqa^2$ and $a$ determines the separation of the phenotypes, the power should depend on $h^2$. The other two parameters $p$ and $d$ are the same as those defined in (3.1.4). Powers with other parameter values can be found in their paper, but the large sample requirement remains unchanged.

The lack of power can be improved if only sibpairs with the extreme values are selected, like the qualitative study with only *affected* sibpairs. Carey and Williamson (1991) proposed the following regression model based on two sibpairs with values $x$ and $y$, where $x$ is the trait value of a specially chosen sib called **proband** and $y$ is the value for one of his/her randomly picked sibings sharing the same environment. Let

$$\pi = (\pi_m - 1)/2,$$
$$\gamma = |\pi| - 1/4,$$

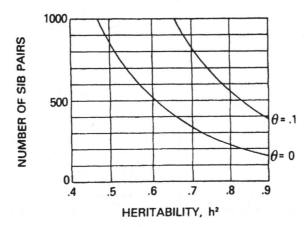

Figure 3.1.1: Sample size for significance level $\alpha = 0.05$, power $= 0.9$, full parental information with 2 marker alleles with equal frequency $p = q = 0.5$, and $\sigma_d^2 = 0$. (From Blackwelder and Elston (1982), copyright © by Marcel Dekker, Inc. Reprinted by permission.)

and

$$y = \beta_1 + \beta_2 x + \beta_3 \pi + \beta_4 \gamma + \varepsilon, \qquad (3.1.16)$$

where $\beta_3$ is related to the additive genetic effect, $\gamma$ to the dominance effect, and the linkage is claimed if $\beta_3 \neq 0$. The reason that $\pi$ instead of $\pi_m$ is used in the regression model (3.1.16) is to make the analysis of (3.1.16) easier, which we will see later. They have found that if the only $x$'s with extreme values are chosen, then the number of sibpairs can be greatly reduced compared with two randomly chosen pairs in the population. Without loss of generality, we consider "extreme" as extremely large in (3.1.2). Fig. 3.1.2 is one of the simulation results they obtained. The parameter $\lambda = 0$ means $d = 0$, i.e., the additive model. The value $q$ is the allele frequency for $B$. Obviously, the larger the $q$, the smaller the power, because large $q$ makes $BB$ dominate the genotypes. In other words, linkage information is reduced due to the second observation, $Y$, is likely to be of the same genotype as the proband. Similar situations happen for other $d$ values (see other examples in Carey and Williamson, 1991).

To see how this method works, we need to add a family index $(i)$ in (3.1.16), i.e.,

$$y_i = \beta_1 + \beta_2 x_i + \beta_3 \pi_i + \beta_4 \gamma_i + \varepsilon_i, \quad i = 1, 2, \ldots, n, \qquad (3.1.17)$$

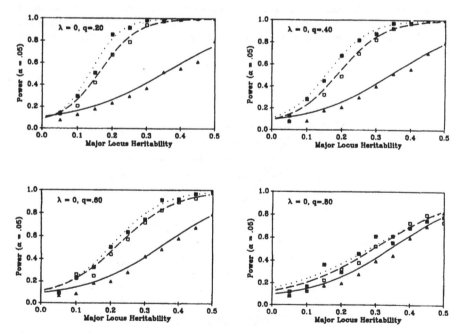

Figure 3.1.2: Power to detect linkage with 240 sibling pairs in a random sample (solid triangles), selected sample when top 10% of the distribution is used as proband (empty squares), and top 5% of the distribution (solid squares). (From Carey and Williamson (1991), copyright © by the University of Chicago Press. Reprinted by permission.)

where $n$ is the number of families. Rewriting (3.1.17) in matrix form, we have

$$Y = X\beta + \varepsilon,$$

where $Y = (y_1, y_2, \ldots, y_n)'$, $\beta = (1, \beta_1, \ldots, \beta_4)'$, $\varepsilon = (\varepsilon_1, \ldots, \varepsilon_n)$, and

$$X = \begin{bmatrix} 1 & x_1 & \pi_1 & \gamma_1 \\ 1 & x_2 & \pi_2 & \gamma_2 \\ \vdots & \vdots & \vdots & \vdots \\ 1 & x_n & \pi_n & \gamma_n \end{bmatrix}.$$

From the elementary regression analysis formula, we know the least squares estimate of $\beta$ is

$$\hat{\beta} = (X'X)^{-1}X'Y.$$

Note that

$$X'X = \begin{bmatrix} n & \Sigma x_i & \Sigma \pi_i & \Sigma \gamma_i \\ - & \Sigma x_i^2 & \Sigma x_i \pi_i & \Sigma x_i \gamma_i \\ - & - & \Sigma \pi_i^2 & \Sigma \pi_i \gamma_i \\ - & - & - & \Sigma \gamma_i^2 \end{bmatrix},$$

where "$-$" are omitted values because $X'X$ is symmetric. Since the other sib is taken at random, $P(\pi = -1/2) = 1/4$, $P(\pi = 0) = 1/2$, and $P(\pi = 1/2) = 1/4$. Hence, for a large sample,

$$\Sigma \pi_i/n \to E[\pi] = (-1/2) \times P(\pi = -1/2) + 0 \times P(\pi = 0) + (1/2) \times P(\pi = 1/2)$$
$$= (-1/2) \times (1/4) + 0 \times 1/2 + (1/2) \times (1/4) = 0.$$

Similarly, we have $\Sigma \gamma_i/n \to 0$ and $\Sigma \pi_i \gamma_i/n \to 0$. Moreover $\Sigma x_i \pi_i/n \to E[X] \cdot E[\pi] \to 0$ and $\Sigma x_i \gamma_i/n \to 0$. Thus, $(X'X)/n$ is approximately

$$(X'X)/n \approx \begin{bmatrix} 1 & \mu_x & 0 & 0 \\ \mu_x & \mu_x^2 + \sigma^2 & 0 & 0 \\ 0 & 0 & 1/8 & 0 \\ 0 & 0 & 0 & 1/16 \end{bmatrix}.$$

Thus,

$$E[\hat{\beta}_3] = 8\Sigma \pi_i y_i/n$$
$$\approx 8\{1/2 \cdot P(\pi = 1/2) \cdot E[y|\pi = 1/2, x \geq x_0] + 0 \cdot P(\pi = 0) \cdot E[y|\pi = 0, x \geq x_0]$$
$$- 1/2 \cdot P(\pi = -1/2) \cdot E[y|\pi = -1/2, x \geq x_0]\}$$
$$= E[y|\pi = 1/2, x \geq x_0] - E[y|\pi = -1/2, x \geq x_0], \tag{3.1.18}$$

where $(x, y)$ are the values for the proband and other sib, and $x_0$ is the threshold for the proband.

Note that if the gene is not linked to the marker, $E[y|\pi = j, x \geq x_0]$ has nothing to do with $j$, and we should have $\hat{\beta}_3 \approx 0$. On the other hand, if they are linked, $E[y|\pi = 1/2, x \geq x_0]$ is expected to be larger than $E[Y|\pi = -1/2, x \geq x_0]$, or $\hat{\beta}_3 > 0$.

Hence testing $\beta_3 = 0$ in (3.1.16) should work. To assess the magnitude of $\beta_3 > 0$ is important in the design stage. Let $\pi_t = IBD$ score of the trait gene,

$$E[y|\pi = i, x \geq x_0] = \sum_{j=0}^{2} E[y|\pi_t = j, x \geq x_0] P[\pi_t = j|\pi = i, x \geq x_0]$$

$$= \sum_{j=0}^{2} E[y|\pi_t = j, x \geq x_0] P[\pi_t = j|\pi = i]. \tag{3.1.19}$$

The values of $P[\pi_t = j|\pi = i]$ can be found in Table 2.3.10 (the $\pi$ here requires a conversion to $\pi_m = 2\pi + 1$ for the $\pi_m$ in Table 2.3.10 because of the new definition in (3.1.16)). If we can find the joint density $f(x, y|\pi_t)$ for $X$ and $Y$, then

$$E[y|\pi_t, x \geq x_0] = \int_{-\infty}^{\infty} \int_{x_0}^{\infty} yf(x, y|\pi_t)dxdy / \int_{-\infty}^{\infty} \int_{x_0}^{\infty} f(x, y|\pi_t)dxdy \tag{3.1.20}$$

can be calculated. Thus, we need to find $f(x, y|\pi_t)$. With random mating assumption, this should be easy with the help of Table 3.1.1. With this table, we need only the density functions of $P(x, y|g_t)$ (see Exercise 3.2). To do this, let $g_t = (g_x, g_y)$ be the genotypes in Table 3.1.1, the mean of phenotype $g_x$ be $\mu + \gamma_x$, and that of $g_y$ be $\mu + \gamma_y$ in (3.1.1) and (3.1.2). Moreover, we decompose $\varepsilon$ in (3.1.1) as $\varepsilon_g = \varepsilon_f + \varepsilon_e$ where $\varepsilon_f$ is the family variance component and $\varepsilon_e$ is the individual variation beyond family. Moreover, we assume $\varepsilon_f$ and $\varepsilon_e$ are independent and normally distributed with $\text{Var}(\varepsilon_f) = \sigma_f^2$ and $\text{Var}(\varepsilon_e) = \sigma_e^2$. Therefore,

$$\sigma_\varepsilon^2 = \sigma_f^2 + \sigma_e^2 \quad \text{and} \quad \sigma^2 = \text{Var}(X - Y) = 2\sigma_e^2.$$

Thus, $P(x, y|g_t)$ has a bivariate normal distribution with mean $(\mu + \gamma_x, \mu + \gamma_y)$ and variance-covariance matrix

$$\Sigma = \begin{bmatrix} \sigma_\varepsilon^2 & \sigma_f^2 \\ \sigma_f^2 & \sigma_\varepsilon^2 \end{bmatrix}.$$

If the means and variances can be specified in the design stage, $E(\hat{\beta}_3)$ in (3.1.18) can be found by working backwards from (3.1.20) to (3.1.18). Due to this lengthy process, Fig. 3.1.2 was obtained by simulation.

The regression equation (3.1.13) or (3.1.15) can be used to test linkage, but not to estimate the distance $\theta$ between the marker and the gene, because $\beta = -(1 - 2\theta)^2 \sigma_a^2$ has two confounding factors, $\theta$ and $\sigma_a^2$. This situation is changed if the suspected gene

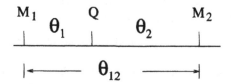

Figure 3.1.3: Gene $Q$ is surrounded by two flanking markers, $M_1$ and $M_2$.

$\theta$ is between two adjacent markers at location $M_1$ and $M_2$. Let the loci and distances $\theta_1, \theta_2, \theta_{12}$ be defined as in Fig. 3.1.3. Since $\theta_{12}$ is known and

$$\theta_1 + \theta_2 - 2\theta_1\theta_2 = \theta_{12}, \qquad (3.1.21)$$

we can estimate $\theta_1$ and $\theta_2$ if we use (3.1.13) or (3.1.15) on both marker loci. More specifically, we can estimate the two $\beta$s in the regression equations (3.1.13) or (3.1.15). They are

$$\beta_1 = -(1 - 2\theta_1)^2\sigma_a^2$$
$$\beta_2 = -(1 - 2\theta_2)^2\sigma_a^2.$$

With the help of (3.1.21), we can solve for $\theta_1$ (and $\theta_2$).

# 3.2 Interval Mapping of Quantitative Trait Loci

As we discussed in §3.1, to locate a quantitative trait locus (QTL) is not easy when applied to human genetics. Environmental variation among a large number of families can cause modeling problems. The situation is different in crop and animal breeding where a large number of offspring are possible from the same parents in the same environment. Moreover, in these instances, mating selection and environment can be strictly controlled. In this section, QTL linkage in crop and animal breeding will be discussed first. Its extension to human gene search is discussed in the last part of this section.

## 3.2.1 Lander and Botstein's Interval Mapping Method

Lander and Botstein's (1989) interval mapping method is one of the most widely used algorithms in gene hunting of QTL for economic crops. In economic crops, such as wheat, tomato, soybean and corn, pure-breed subjects are always available either from previous genetic experiment or by consecutive backcrossing. We consider only two pure breeds with homogeneous markers. Suppose the trait is affected by only one gene and the two pure lines were mated as in Fig. 3.2.1.

In this figure, $M_1$, $M_2$ and $M_3$ are markers with known locations and Q, the gene we wish to locate, may or may not be there. The $\theta_1$ and $\theta_2$ are the distances (recombination fractions, to be exact) between Q and $M_1$ and $M_2$. The sum $\theta_1 + \theta_2 - 2\theta_1\theta_2 \equiv \theta_{12} \equiv r$ is the distance between $M_1$ and $M_2$. For small $r$, we may neglect the cross-product term and use $\theta_1 + \theta_2 = r$. The allele $g_2$ in $F_2$ can be $B$ or $b$, depending on the crossover in the meiosis. The sampling scheme is at $F_2$. Thus, the data in the $F_0$ and $F_1$ may be ignored in data analysis. By assuming an additive model, we let the phenotype of $B$ and $b$ be

$$y_{BB} \sim N(\mu_{BB}, \sigma^2), \quad \text{and} \quad y_{Bb} \sim N(\mu_{Bb}, \sigma^2),$$

|  | $M_1$ | $\theta_1$ | $Q$ | $\theta_2$ | $M_2$ |  | $M_3$ |  |  | $M_1$ | $\theta_1$ | $Q$ | $\theta_2$ | $M_2$ |  | $M_3$ |
|---|---|---|---|---|---|---|---|---|---|---|---|---|---|---|---|---|
| $F_0$ | $-\perp-$ | | $\perp-$ | | $\perp-$ | | $\perp$ | $\times$ | | $\perp-$ | | $\perp-$ | | $\perp-$ | | $\perp$ |
| | 1 | $B$ | 1 | | 1 | | | | 2 | | $b$ | 2 | | | 2 | |
| | 1 | $B$ | 1 | | 1 | | | | 2 | | $b$ | 2 | | | 2 | |

|  | $M_1$ | $\theta_1$ | $Q$ | $\theta_2$ | $M_2$ | $M_3$ |
|---|---|---|---|---|---|---|
| $F_1$ | $-\perp-$ | | $\perp-$ | | $\perp-$ | $\perp-$ |
| | 1 | $B$ | 1 | | 1 | |
| | 2 | $b$ | 2 | | 2 | |

$F_2 = F_1 \times BB$

|  | $M_1$ | $\theta_1$ | $Q$ | $\theta_2$ | $M_2$ | $M_3$ |
|---|---|---|---|---|---|---|
| CASE 1 | $-\perp-$ | | $\perp-$ | | $\perp-$ | $\perp-$ |
| | 1 | | $B$ | | 1 | |
| | 1 | | $g_2$ | | 1 | |
| CASE 2 | 1 | | $B$ | | 1 | |
| | 2 | | $g_2$ | | 2 | |
| CASE 3 | 1 | | $B$ | | 1 | |
| | 1 | | $g_2$ | | 2 | |
| CASE 4 | 1 | | $B$ | | 1 | |
| | 2 | | $g_2$ | | 1 | |

Figure 3.2.1: Schematic diagram for backcross breeding of pure brands.

or, without loss of generality, we may let

$$y_i = \mu + a\xi_i + e_i, \tag{3.2.1}$$

where $\xi_i = 1$ if the genotype is $BB$ and $-1$ if the genotype is $Bb$. Given the marker conditions, the probability of $g_2$ being $B$ or $b$ can be summarized as in Table 3.2.1 (the subject index $i$ is omitted).

If we define the marker condition for plant $i$ to be $m_i$, i.e., $m_i = (11/11)$, $(12/12)$, $(11/12)$, or $(12/11)$ in Table 3.2.1, and $\Pr\{g_2 = B|m_i\} = p_{m_i}$, then the density function of $y_i$ becomes

$$f(y_i|m_i) = p_{m_i}\phi(\mu + a, \sigma^2) + (1 - p_{m_i})\phi(\mu - a, \sigma^2), \tag{3.2.2}$$

where $\phi(\mu, \sigma^2)$ denotes the density function of $N(\mu, \sigma^2)$. The likelihood function becomes

$$L(\mu, a, \sigma^2|\{y_i,\ m_i\}) = \prod_{i=1}^{n} f(y_i|m_i). \tag{3.2.3}$$

The LOD score with given $\theta_1$ for testing $\beta \neq 0$ becomes

$$\text{LOD} = \log_{10} L(\hat{\mu}^*, \hat{a}^*, \hat{\sigma}^2) - \log_{10} L(\tilde{\mu}, \tilde{\sigma}^2), \tag{3.2.4}$$

Table 3.2.1: Chance of $g_2$ being $B$ in $F_2$ given the markers' conditions.

| Marker | | $(m)$ | $p_m \equiv Pr\{g_2 = B|m\}$ | For small $\theta_1, \theta_2$ |
|--------|--------|-------|------------------------------|-------------------------------|
| $M_1$ | $M_2$ | | | |
| 11 | 11 | (11/11) | $(1-\theta_1)(1-\theta_2)/[(1-\theta_1)(1-\theta_2) + \theta_1\theta_2]$ | 1 |
| 12 | 12 | (12/12) | $\theta_1\theta_2/[(1-\theta_1)(1-\theta_2) + \theta_1\theta_2]$ | 0 |
| 11 | 12 | (11/12) | $(1-\theta_1)\theta_2/[(1-\theta_1)\theta_2 + \theta_1(1-\theta_2)]$ | $\theta_2/(\theta_1 + \theta_2)$ |
| 12 | 11 | (12/11) | $(1-\theta_2)\theta_1/[(1-\theta_1)\theta_2 + \theta_1(1-\theta_2)]$ | $\theta_1/(\theta_1 + \theta_2)$ |

where $\hat{\mu}^*, \hat{a}^*, \hat{\sigma}^2$ are the mle based on the model (3.2.1) and $\tilde{\mu}$, $\tilde{\sigma}^2$ are the mle of (3.2.1) under $H_0$: $a = 0$. The distribution of the LOD defined in (3.2.4) is not easy to obtain because $\theta_1$ is unknown in practice. It will be soon clear why $\theta_1$ can be considered as known. Moreover, it will also be clear that the exact distribution of (3.2.4) is not important. However, 4.6LOD should have roughly a chi-square distribution with 1 degree of freedom. The mle of (3.2.3), though it has no neat form because it is a mixture of two normal distributions, can easily be found numerically. Lander and Botstein have a program, MAPMAKER-QTL, that is available upon request ( http://www-genome.wi.mit.edu/genome_software ).

With given marker information, the LOD score at any given location can be found. A locus with a high LOD score is a good indicator a QTL is located there. Fig. 3.2.2 demonstrates the procedure on how a $LOD(x)$ curve can be obtained for each chromosome. In this figure, $x$ represents the location measure from one end of the chromosome, and $M_1, M_2, \ldots$, are locations of the markers with their distances $x_1, x_2, \ldots$ measured from the tip of the chromosome. The LOD score at point $x$ is based on $M_2$ and $M_3$ with $\theta_1 \approx x - x_2$ and $\theta_2 \approx x_3 - x$. Thus, we can assume that $\theta_1$ is known in

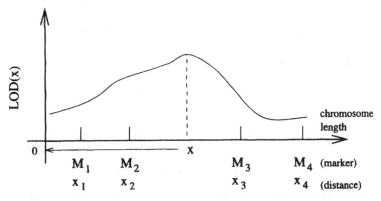

Figure 3.2.2: LOD score for QTL on a dense-marker map. x, sliding along the x-axis, is supposedly the (a) gene's position.

(3.2.4). Since the distance and recombination fraction are not the same for long distances, the Haldane transform (see 1.2.1) is used to convert distance into a recombination fraction, i.e.,

$$\theta_1 = \frac{1}{2}(1 - e^{-2|x-x_2|}), \quad \text{and} \quad \theta_2 = \frac{1}{2}(1 - e^{-2|x-x_3|}). \tag{3.2.5}$$

Because $x$ is evaluated at a large number of points, locus identification is now a very high dimensional multiple comparisons problem. That is why we are not interested in the exact distribution of $LOD(x)$. Lander and Botstein simulated the appropriate thresholds for various chromosome at different marker densities with a testing significance level of 0.05 (Fig. 3.2.3). Note that the values in the x-axis of Fig. 3.2.3 are the spacing between markers in the x-axis of Fig. 3.2.2. Each chromosome is assumed to be 100 cM in length. For example, a LOD threshold of 2.4 is required when using a 15 cM interval map in a 1000 cM geneome (10 chr curve in the graph).

They also did a simulation study on the feasibility of applying their method to multiple genes. They assumed that there were 12 100 cM chromosomes with 5 QTLs that jointly affected a quantitative trait. Each of the first 5 chromosomes had one of the genes. The genes' contributions to the trait were, respectively, $|\mu_{BB} - \mu_{Bb}|/\sigma = 1.5$,

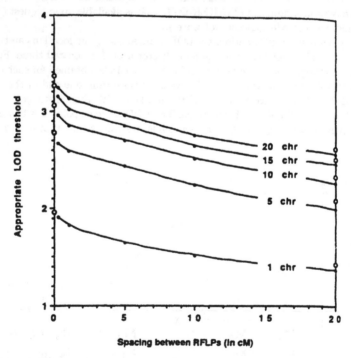

Figure 3.2.3: LOD thresholds for false-alarm rate at 0.05 anywhere in the genome. (From Lander and Botstein(1989), copyright ©by The Genetics Society of America. Reprinted by permission.)

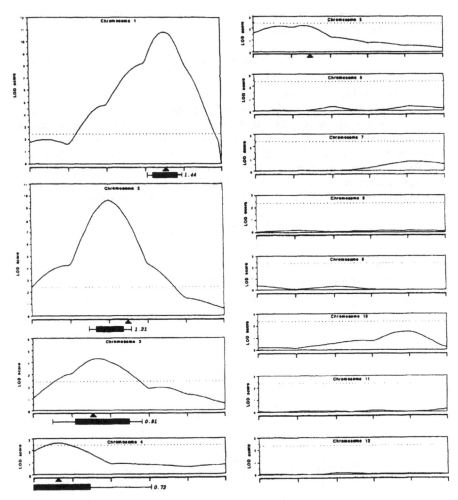

Figure 3.2.4: LOD score of a sliding gene loci in Lander and Botstein's simulation experiment. The length of each chromosome is 100 cM. The RFLP markers are spaced every 20 cM indicated by tick marks. The QTL positions are indicated by black triangles. The dotted lines are at LOD = 2.4, the required 0.05 significance level threshold as suggested by Figure 3.2.3. (From Lander and Botstein (1989), copyright © by The Genetics Society of America. Reprinted by permission.)

1.25, 1.0, 0.75 and 0.5 with additive effects. The number of $F_2$ backcross progeny in the simulation was 250. The result, shown in Fig. 3.2.4, appears very successful.

It is interesting to examine the advantage of locating QTL using a dense marker map. Since QTL identification based on LOD is equivalent to testing of a regression model, we may use the expected mean square (EMS) to compare the efficiency with various marker information (see Appendix A.2). It is known that the EMS of the main effect in one-way analysis of variance is

$$\delta^2 = \sum_{j=1}^{k} n_j(\mu_j - \bar{\mu})^2/\sigma^2, \qquad (3.2.6)$$

where $k$ is the number of categories, $n_j$ and $\mu_j$ are the number of observations and true means in the $j$th category, $\bar{\mu} = \sum_{j=1}^{k} \mu_j/k$, and $\sigma^2$ is the error variance for $\varepsilon_i$ in (3.2.1) (see A.2.8 and A.2.10). Since (3.2.6) shows how the sample size affects the power of a test, we may use it to measure the efficiency between tests. For example, if Test 1 requires sample size $2n$ and Test 2 requires only $n$ to reach the same $\delta^2$, then we say that Test 1 is only half as efficient as Test 2. In practice, the number of offspring in any genotype category $m$ in Table 3.2.1 is a random variable. Let there be $N_j$ offsprings with marker $j$ in the four $m$ categories in Table 3.2.1. Then the expected EMS is

$$E[\delta^2] = \sum_{j=1}^{4} E[N_j](\mu_j - \bar{\mu})^2/\sigma^2, \qquad (3.2.7)$$

where $j$ runs from the top to bottom in Table 3.2.1. Let the total number of offspring be $n$. For a small distance, or as the first-order approximation,

$$E[N_1] = E[N_2] = n(1 - \theta_1 - \theta_2)/2; \quad E[N_3] = E[N_4] = n(\theta_1 + \theta_2)/2.$$

Their means, by neglecting double crossing, are

$$E[\mu_1] = \mu_{BB}, \quad E[\mu_2] = \mu_{Bb},$$

$$E[\mu_3] = \frac{\theta_2}{\theta_1 + \theta_2}\mu_{BB} + \frac{\theta_1}{\theta_1 + \theta_2}\mu_{Bb}$$

$$E[\mu_4] = \frac{\theta_1}{\theta_1 + \theta_2}\mu_{BB} + \frac{\theta_2}{\theta_1 + \theta_2}\mu_{Bb}.$$

Using algebraic manipulation, we have

$$\delta^2 = n\left[(1 - r) + \frac{(\theta_2 - \theta_1)^2}{r}\right](\mu_{BB} - \mu_{Bb})^2/(4\sigma^2).$$

where $r \approx \theta_1 + \theta_2$ is the recombination fraction between the two flanking markers. The worst case is when $\theta_1 = \theta_2 = r/2$, or

$$\delta^2_{\min} = n(1 - r)(\mu_{BB} - \mu_{Bb})^2/(4\sigma^2) = n(1 - 2\theta_1)(\mu_{BB} - \mu_{Bb})^2/(4\sigma^2). \qquad (3.2.8)$$

Thus, the increase in efficiency is proportional to $(1 - r)$, or, more precisely, $(1 - r^*)$, where $r^*$ is the recombination fraction of $r$ converted by Haldane's transform (3.2.5).

When very sparse markers are available, multiple marker gene search is equivalent to use one marker at a time. Suppose in Fig. 3.2.1 only marker $M_1$ is used. In this case, it can be seen that the mean values for an offspring with markers 11 or 12 are

$$\mu_{11} = (1 - \theta_1)\mu_{BB} + \theta_1\mu_{Bb}, \quad \text{and} \quad \mu_{12} = (1 - \theta_1)\mu_{Bb} + \theta_1\mu_{BB}, \quad \text{and}$$
$$\delta^2 = n(1 - 2\theta_1)^2(\mu_{BB} - \mu_{Bb})^2/(4\sigma^2). \qquad (3.2.9)$$

Comparing (3.2.8) and (3.2.9) we see the loss of efficiency due to single markers is approximately $(1 - 2\theta_1)$.

## 3.2.2  Intercross Experiments for Nonadditive Effects

As we have discussed in §3.1, there may be more than just the additive effect between the alleles $B$ and $b$. Dominance effect may exist. The backcross experiment in Fig. 3.2.1 cannot discover this fact because there is no genotype $bb$. If we let the $F_2$ in Fig. 3.2.1 be an intercross experiment, then all the genotypes, $BB(g \equiv 1)$, $Bb(g \equiv 2)$, and $bb(g \equiv 3)$, appear in $F_2$. Using the notation in (3.1.2) and (3.2.2), we have

$$f(y_i|m_i) = \sum_{g=1}^{3} \phi(\gamma_g, \sigma^2) P(g|m_i),  \tag{3.2.10}$$

where $p(g|m_i)$ is the conditional probability of having genotype $g$ in the gene locus given the marker information $m_i$ for individual $i$. The conditional probabilities $p(g|m_i)$ can be obtained by using the gamete frequencies for the parents in $F_1$ in meiosis. These are given in Table 3.2.2. Using this table, we can compute all $p(g|m_i)$, e.g.,

$$P(BB|11/11) = \frac{(1 - \theta_1)^2 (1 - \theta_2)^2/4}{\left[(1 - \theta_1)^2(1 - \theta_2)^2 + 2\theta_1\theta_2(1 - \theta_1)(1 - \theta_2) + (\theta_1\theta_2)^2\right]/4}$$
$$= (1 - \theta_1)^2(1 - \theta_2)^2/(1 - r)^2,$$

$P(Bb|11/11) = 2\theta_1\theta_2(1 - \theta_1)(1 - \theta_2)/(1 - r)^2$, and $P(bb|11/11) = (\theta_1\theta_2)^2/(1 - r)^2$, where $r = \theta_1 + \theta_2 - 2\theta_1\theta_2$.

After being modified for intercross experiment, the backcross likelihood equation (3.2.3) becomes

$$L(\mu, a, d, \sigma^2|\{y_i, \ m_i\}) = \prod_{i=1}^{n} f(y_i|m_i),  \tag{3.2.11}$$

with the new density function $f$ defined by (3.2.10). The LOD equation (3.2.4) becomes

$$\text{LOD} = \log_{10} L(\hat{\mu}^*, \hat{a}^*, \hat{d}^*, \hat{\sigma}^2) - \log_{10} L(\tilde{\mu}, \tilde{\sigma}^2),  \tag{3.2.12}$$

where $\hat{\mu}^*, \hat{a}^*, \hat{d}^*, \hat{\sigma}^2$ are the mle based on the model (3.1.2) and $\tilde{\mu}$, $\tilde{\sigma}^2$ are the mle of (3.1.2) under $H_0$: $a = d = 0$.

Table 3.2.2: Gamete frequencies by $F_1$ in
Fig. 3.2.1 after meiosis.

| Haploid | Probability |
|---------|-------------|
| 1B1 or 2b2 | $(1 - \theta_1)(1 - \theta_2)/2$ |
| 1B2 or 2b1 | $(1 - \theta_1)\theta_2/2$ |
| 2B1 or 1b2 | $\theta_1(1 - \theta_2)/2$ |
| 1b1 or 2b2 | $\theta_1\theta_2/2$ |

Table 3.2.3: Conditional expection $E(Y|m)$ in $F_2$ of an intercross experiment.

| Marker type $m$ | $E(Y|m)$ |
|---|---|
| 11/11 | $\mu + \{a[(1 - \theta_1)^2(1 - \theta_2)^2 - (\theta_1\theta_2)^2] + 2d\theta_1\theta_2[(1 - \theta_1)(1 - \theta_2)]\}/(1 - r)^2$ |
| 11/12 | $\mu + \{a[\theta_2(1 - \theta_2)((1 - \theta_1)^2 - \theta_1^2)] + d[\theta_1(1 - \theta_1)[(1 - \theta_2)^2 + \theta_2^2]\}/[r(1 - r)]$ |
| 11/22 | $\mu + \{a[\theta_2^2(1 - \theta_1)^2 - \theta_1^2(1 - \theta_2)^2] + 2d\theta_1\theta_2[(1 - \theta_1)(1 - \theta_2)]\}/r^2$ |
| 12/12 | $\mu + d[(1 - \theta_1)^2 + \theta_1^2][(1 - \theta_2)^2 + \theta_2^2]/[r^2 + (1 - r)^2]$ |

Sliding the $x$ along the chromosome as it is in Fig. 3.2.2, similar LOD curves can be obtained. Again the threshold is not easy to determine in this sliding process and we may use Fig. 3.2.3 as a guideline.

Maximizing (3.2.11) has to be done numerically even when $\theta_1$ and $\theta_2$ are given. We may use the ordinary least squares method to find the unknown parameters $\hat{\mu}^*$, $\hat{a}^*$, $\hat{r}^*$ by the method of moments as suggested by Haley and Knott (1992). From the previous computation of $P(BB|11/11)$, $P(Bb|11/11)$, and $P(bb|11/11)$,

$$E(y_i|11/11) = \mu + \{a[(1 - \theta_1)^2(1 - \theta_2)^2 - (\theta_1\theta_2)^2] + 2d\theta_1\theta_2[(1 - \theta_1)(1 - \theta_2)]\}/(1 - r)^2.$$

Table 3.2.3 gives the conditional expectation for four marker types. The other five types, 22/22, 22/12, 22/11, 12/11, 12/22 can be obtained by symmetry.

For given $\theta_1$ and $\theta_2$, the equation

$$y_i = E(y|m_i) + \varepsilon_i \tag{3.2.13}$$

is linear in $\mu$, $a$ and $d$. They can be estimated easily because least squares estimates are available in almost all statistical packages.

### 3.2.3   Extension to Sibpair Study

Lander and Botstein's method can be extended to human gene search with sibpairs. Recall in §3.1, if we let $\Delta^2$ be the square of the trait-value difference between the two sibs, then

$$E(\Delta^2|\pi_m = i) = \sigma^2 + 2(1 - 2\theta(1 - \theta))\sigma_a^2 - (1 - 2\theta)^2\sigma_a^2 i.$$

In particular, if $\theta = 0$, then $\pi_m = \pi_t =$ IBD at the gene and

$$E(\Delta^2|\pi_t = i) = \sigma^2 + 2\sigma_a^2 - \sigma_a^2 i \tag{3.2.14}$$

Thus, to make inference on the additive genetic trait, we can use the regression equation

$$\Delta^2 = \alpha + \beta\pi_t + \varepsilon. \tag{3.2.15}$$

Although $\pi_t$ is unknown, we can estimate it when its position is assumed to be known by following the idea of an exhaustive search scheme in the interval mapping

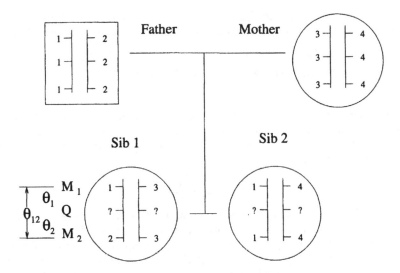

Figure 3.2.5: The relation between siblings' IBD and their parents' genotypes.

method (see Fig. 3.2.2). Since the position of the gene is now "known," we can use the marker information to estimate $\pi_t$. Using the extremely polymorphic markers in Fig. 3.2.5 as an example, the chance that the sibs share the gene at the father's side $\theta_2/(\theta_1 + \theta_2)$ and at the mother's side is 0, assuming that the chance of double cross-over between the two markers is negligible. Thus,

$$\hat{\pi}_t = \theta_2/(\theta_1 + \theta_2). \tag{3.2.16}$$

For less polymorphic markers, to estimate $\pi_t$ can be quite tedious. Fulker and Cardon (1994) and Fulker, Cherny and Cardon (1995) suggest using the regression equation

$$\pi_t = \gamma_0 + \gamma_1 \pi_1 + \gamma_2 \pi_2 + \varepsilon \tag{3.2.17}$$

to estimate $\pi_t$ by the IBD numbers $\pi_1$ and $\pi_2$ at the two markers, respectively. Since we know how to estimate $\pi_1$ and $\pi_2$ for non-polymorphic markers (see 3.1.14), we can thus estimate $\pi_t$ by (3.2.17) if $\gamma_0, \gamma_1$ and $\gamma_2$ are known. The solution of $\gamma_1$ and $\gamma_2$ follows the standard regression normal equation

$$\begin{bmatrix} \text{Var}(\pi_1) & \text{Cov}(\pi_1, \pi_2) \\ \text{Cov}(\pi_1, \pi_2) & \text{Var}(\pi_2) \end{bmatrix} \begin{bmatrix} \gamma_1 \\ \gamma_2 \end{bmatrix} = \begin{bmatrix} \text{Cov}(\pi_t, \pi_1) \\ \text{Cov}(\pi_t, \pi_2) \end{bmatrix}. \tag{3.2.18}$$

All the variances and covariance in (3.2.18) are easy to compute. Actually, we only need to compute one variance and one covariance and all the rest follow just by changing symbols. Obviously, $E(\pi_1) = 1$, $\text{Var}(\pi_1) = 1/2$. Moreover,

$$\text{Cov}(\pi_1, \pi_2) = (1 - 2\theta_{12})^2/2. \tag{3.2.19}$$

Similarly, $\text{Cov}(\pi_t, \pi_i) = (1 - 2\theta_i)^2/2$ for $i = 1, 2$. The derivation of (3.2.19) is left for the exercise. It can be shown (Elston and Keats, 1985) by solving (3.2.18), that

$$\hat{\gamma}_1 = [(1 - 2\theta_1)^2 - (1 - 2\theta_2)^2(1 - 2\theta_{12})^2]/[1 - (1 - 2\theta_{12})^4],$$

$$\hat{\gamma}_2 = [(1 - \theta_2)^2 - (1 - 2\theta_1)^2(1 - 2\theta_{12})^2]/[1 - (1 - 2\theta_{12})^4], \qquad (3.2.20)$$

$$\hat{\gamma}_0 = 1 - \hat{\gamma}_1 - \hat{\gamma}_2.$$

The last equation becomes obvious if we take the expectation on both sides of (3.2.17). Once $\pi_t$ has been estimated for all the sibpairs at a given location $x$ in Fig. 3.2.2, we can find the $\hat{\beta}(x)$ in (3.2.15). The most significant value of $\beta(x)$, measured by the usual $t$-statistic in regression,

$$t(x) = \frac{-\hat{\beta}(x)}{SE(\hat{\beta}(x))}$$

is used to select $x$, where $SE(\hat{\beta}(x))$ is the standard error of $\hat{\beta}$ in regression analysis. Note that this $t(x)$ may not have a $t$-distribution due to the fact that $\varepsilon$ in (3.2.15) is not normally distributed. Moreover, this is a multiple comparisons problem. The exact threshold of accepting $\beta(x) < 0$ at a given significance level is not easy to derive. There seems to be no definite threshold except to claim that $\alpha = 0.001$ for the $t$-distribution (roughly at LOD = 3.0 level) should be enough to claim linkage. The simulation result for a 100 cM chromosome with 20 cM and 10 cM marker intervals given in Fig. 3.2.6 indicates that this method should work well.

Cardon and Fulker (1994) have extended their results from two randomly selected sibpairs to two sibs with extreme values similar to what was discussed at the end of §3.1. Recall that in a proband model

$$y = \beta_1 + \beta_2 x + \beta_3 \pi + \beta_4 \gamma + \varepsilon, \qquad (3.2.21)$$

where all the symbols are defined in (3.1.16), and the only difference is now

$$\pi = (\hat{\pi}_t - 1)/2, \quad \gamma = |\pi| - 1/4$$

and $\pi_t$ is estimated using (3.2.17). Fig. 3.2.7 shows one of their simulation result. Note that the power rises from single marker 0.28 to 0.57 for 10% proband to 0.65 for 5% and 0.84 for 1%.

Detecting linkage by (3.2.15) is based on the method of moments because it uses the expectations. Maximum likelihood type of estimation can also be developed. Let $I_m(j)$ denote the marker information of the two sibs in family $j$, with $j = 1, 2, \ldots, n$ and $\Delta_j^2 = (Y_{1j} - Y_{2j})^2$, where $Y_{1j}$ and $Y_{2j}$ denote the phenotypes of the two siblings. Thus, the likelihood for family $j$ is

$$L_j = P(\Delta_j^2 \cap I_m(j)) = \sum_{i=0}^{2} P(\Delta_j^2 \cap I_m(j)|\pi_t = i)P(\pi_t = i)$$

$$= \sum_{i=0}^{2} P(\Delta_j^2|\pi_t = i)P(I_m(j)|\pi_t = i)P(\pi_t = i)$$

$$= P(I_m(j)) \sum_{i=0}^{2} P(\Delta_j^2|\pi_t = i) \cdot P(\pi_t = i|I_m(j)). \qquad (3.2.22)$$

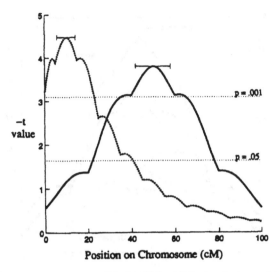

Position on Chromosome (cM)

Figure 3.2.6: The t-statistics curves produced by 1000 simulations of 1000 sibpairs for a 100 cM chromosome segment. The solid line shows the results for a single QTL located at 50 cM and flanked by successive markers spaced at 20 cM intervals. The dotted line represents a single QTL located at 10 cM and flanked by markers spaced 10 cM apart. The QTL heritability is set at 0.5 and all the markers have six alleles with equal frequency. The horizontal error bars describe the 95% confidence interval for the estimate of QTL position. The two horizontal dotted lines indicate the 5% and 0.1% significance thresholds for a normal distribution. (From Fulker and Cardon (1994). copyright © by The University of Chicago Press. Reprinted by permission.)

Note that the third equality follows from the independence of $\Delta_j^2$ and $I_m(j)$ given $\pi_t$. Kruglyak and Lander (1995) assumed that $P(\Delta_j^2|\pi_t = i)$ in density form,

$$f(\Delta_j^2|\pi_t = i) = \frac{1}{\sqrt{2\pi\sigma_i^2}}\exp(-\frac{\Delta_j^2}{2\sigma_i^2}), \ i = 0, 1, 2, \tag{3.2.23}$$

and $\sigma_0^2 > \sigma_1^2 > \sigma_2^2$ if the gene at locus $t$ affects the phenotype, and $\sigma_0^2 = \sigma_1^2 = \sigma_2^2$ under $H_0$ that locus $t$ is irrelevant to the trait value. Note that $E\Delta_i = 0$ because the order of the two sibs is arbitrary. They called this the **mle variance method** and named a special case the **no dominance variance model** if the restriction $\sigma_1^2 = (\sigma_0^2 + \sigma_2^2)/2$ is imposed. Thus, we can do a likelihood ratio test or compute LOD score by the likelihood equation

$$L = \prod_{j=1}^{n} L_j \tag{3.2.24}$$

if the other two terms in (3.2.22) can be computed. One can see that $P(I_m(j))$ plays no role in the likelihood ratio test. For any assumed gene location using the interval mapping method, $P(\pi_t = i|I_m(j))$ can be found following the evaluation of (3.2.17).

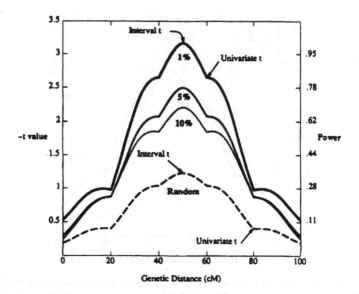

Figure 3.2.7: The average $t$-statistic curves produced by 500 simulations of 250 sib-pairs for a 100 cM chromosome segment. The QTL, with heritability 0.30, is located at 50 cM and is flanked by successive markers having 10 alleles at 20 cM intervals. Solid lines correspond to 10%, 5% and 1% selection on the lower tail of a random sample distribution, while the dashed line represents a random sample. The $t$-statistics and power values at $\alpha = 0.05$ for the marker positions correspond to univariate sibpair regressions. The interval-$t$ corresponds to (3.2.21) when $\pi$ is estimated by $\hat{\pi}_t$, while univariate-$t$ corresponds to $t$ at the marker positions. The two $t$-values coincide at the marker loci. (From Cardon and Fulker (1994) copyright © by The University of Chicago Press. Reprinted by permission.)

Thus, (3.2.22) is completely specified for any $\pi_t(\theta)$. Again, the threshold to claim significance based on maximum of the LOD scores using a sliding window is difficult to determine analytically. Further simulation is necessary.

The LOD score can also be computed for the original Haseman-Elston (H-E) method (3.1.13), i.e., to approximate the distribution of $\Delta_j^2$ by the regression likelihood equation,

$$f(\Delta_j^2|\pi_t \doteq i) = \frac{1}{\sqrt{2\pi}\sigma}\exp(\Delta_j^2 - \alpha - \beta i)^2/(2\sigma^2)], \qquad (3.2.25)$$

under linkage hypothesis and $\beta = 0$ under $H_0$.

Both (3.2.23) and (3.2.25) are parametric models depending on the normality approximation. Kruglyak and Lander (1995) suggested a nonparametric method by using the ranks of $\{\Delta_j^2\}_{j=1}^n$. Let the rank $r_j$, with possible values from the smallest 1 to the highest $n$, be the rank of $\Delta_j^2$. Define a test statistic

$$X = \sum_{j=1}^{n} r_j f(\pi_{tj}), \qquad (3.2.26)$$

where $f()$ is a monotonic function of the IBD number $\pi_{tj}$ of the $j$th sibpair. Obviously, if $f()$ is a monotonic decreasing function, then the larger the $X$, the more likely that there is a linkage between the trait and marker loci. Kruglyak and Lander suggested an intuitive function

$$f(0) = 1, \quad f(1) = 0, \quad f(2) = -1.$$

Under the null hypothesis, $X$ should have a distribution of

$$X_0 = \sum_{j=1}^{n} r_j f(\pi_{0j}),$$

where $(\pi_{01}, \pi_{02}, \ldots, \pi_{0n})$ is a random permutation of $(\pi_{t1}, \pi_{t2}, \ldots, \pi_{tn})$, because in absence of linkage, these ranks have equal chance to be attached to any $\pi_{tj}$. The mean $E[Y_0]$ and variance $V[X_0]$ of $X_0$ can be found by exhaustive listing (small $n$) or simulation. The Z-score of the observed $X$ is then

$$Z = \frac{X - E[X_0]}{\sqrt{V[X_0]}}. \qquad (3.2.27)$$

Kruglgak and Lander suggest using $Z > 4.1$ for claiming linkage at $\alpha = 0.05$ level of significance for a genomewide search. A computer program, MAPMARKER/SIBS, to do a test based on (3.2.23), (3.2.25) and (3.2.27), can be found in Appendix B.6.

The following numerical example may help our understanding of the outputs from MAPMARKER/SIBS (see Appendix B.6).

**Example 3.2.1** *Find the LOD scores by the H-E method and mle variance method and the Z-score by the nonparametric method.*

Pedigree information for six sibpairs.

| Family | Parents markers | sib 1 | (type and QT value) | sib 2 | (type and QT value) |
|--------|-----------------|-------|---------------------|-------|---------------------|
| 1 | 13 × 12 | 13 | 0.0 | 12 | 3.0 |
| 2 | 13 × 12 | 11 | 1.0 | 12 | 2.0 |
| 3 | 13 × 12 | 13 | 0.0 | 12 | 2.0 |
| 4 | 13 × 12 | 11 | 0.0 | 12 | 1.0 |
| 5 | 13 × 12 | 12 | 1.0 | 12 | 1.0 |
| 6 | 13 × 12 | 13 | 2.0 | 13 | 3.0 |

**Sol**: From this table, we can find $\pi_{tj} = $ IBD for the $j$th sibpair, $\Delta_j^2$ and $r_j$ as follows.

| $j$ | 1 | 2 | 3 | 4 | 5 | 6 |
|---|---|---|---|---|---|---|
| $\pi_{tj}$ | 0 | 1 | 0 | 1 | 2 | 2 |
| $\Delta_j^2$ | 9 | 1 | 4 | 1 | 0 | 1 |
| $r_j$ | 6 | 3 | 5 | 3 | 1 | 3 |

The LOD score for Haseman-Elston test in

$$LOD_{HE} = \log_{10}\left[\prod_{j=1}^{n}(2\pi\hat{\sigma}_1)^{-1}\exp(\Delta_j^2 - \hat{\alpha} - \hat{\beta}\pi_{tj}]^2/2\hat{\sigma}_1^2\right]$$

$$- \log_{10}\left[\prod_{j=1}^{n}(2\pi\hat{\sigma}_0)\exp(\Delta_j^2 - \hat{\mu})^2/\hat{\sigma}_0^2\right],$$

$$= -\frac{n}{2}\left[\log_{10}\hat{\sigma}_1^2 - \log_{10}\hat{\sigma}_0^2\right],$$

where the head represents the *mle* estimates of the parameters. By simple computation, $\hat{\sigma}_1^2$ = residual sum of squares/6 = 3.55, $\sigma_0^2$ = (sample variance) $\times(5/6)$ = 9.55 and $LOD_{HE}$ = 1.289.

The LOD score based on (3.2.23) can be computed by noting $\hat{\sigma}_0^2 = (9+4)/2 = 6.5, \hat{\sigma}_1^2 = (1+1)/2 = 1$ and $\hat{\sigma}_2^2 = (1+0)/2 = 0.5$, and the variance under $H_0$ is $\hat{\sigma}^2 = (9+4+1+1+1+0)/6 = 2.667$. Thus,

$$LOD_{mlev} = \frac{6}{2}\log_{10}(2.667) - \frac{2}{2}[\log_{10}(6.5) + \log_{10}(1) + \log_{10}(0.5)]$$

$$= 0.768$$

For the nonparimatric Z-score, by exhaustive listing, we have $E[X_0] = 0$ and $var[X_0] = 12.4$. Since $X = 6 + 5 - 1 - 3 = 7$, $Z = 7/\sqrt{12.4} = 1.988$. $\square$

### 3.2.4 Examples in Real Data Analyses

Table 3.2.4 shows a few examples when Lander and Bostein's method was used in practice.

### 3.2.5 Polygene Detection by Dense Markers

As we have seen from the simulation study in Fig. 3.2.4 and from a real application in Fig. 3.2.8, interval mapping can detect multiple genes. However, in interval mapping, only two flanking loci are considered for each candidate gene. It is well known in multivariate or regression analysis that if a response $y$ is caused by many variables, such as in the cases of multiple genes, it is more appropriate to study the partial correlation coefficient so that the effects of one particular variable (gene) can be isolated (see Appendix A.1). Zeng (1993, 1994) and Jiang and Zeng (1995) pointed out

Table 3.2.4: Examples in QTL gene search.

| Source | Paterson et al. (1988) | Keim et al. (1990) | Rise (1991) | Asada et al. (1994) |
|---|---|---|---|---|
| Species | Tomato | Soybean | Mouse | Mouse |
| # Chromosome | 12 | 26 | 19 | 1 Examined |
| Total markers | 300 | 128 | 67 | 12 |
| Sample size | 237 | 60 | 251 | 67 |
| | Backcross | Backcross | Backcross | Intercross |
| Trait of | Fruit mass, | leaf, flower, | seizure | testicular |
| interest | pH-value | seed properties | frequency | teratomas |
| Findings | see Fig. 3.2.8 | [a] | [b] | see Fig. 3.2.9 |

[a] Marker loci related to leaf width, leaf length, stem diameter, stem length, canopy height, first flower, seed pod maturity, and seed-fill are found.
[b] Marker loci related to seizure was found on chromosome 9 with $LOD = 5.88$.

how dense-marker information can be used in polygene detection. (A polygene is one of the multiple genes that affect a given trait.) For simplicity, we follow their derivation for the backcrossing model on homozygous $F_0$ as shown in Fig. 3.2.1.

Suppose trait $y$ is affected by $m$ gene loci. Let the genotypes of the two pure-bred $F_0$ parents be denoted as $BB$ and $bb$ for all the genes. Let the mean contributions from the $u$th gene be $\mu_{BB}(u)$ and $\mu_{bb}(u)$ for the genotypes $BB$ and $bb$ respectively, with $u = 1, 2, \ldots, m$. Define $c_u = \mu_{BB}(u) - \mu_{bb}(u)$, where $c_u$ can be positive or negative. Assuming no epistasis, the mean difference between the two pure-bred parents can be described by

$$\Delta\mu = \sum_{u=1}^{m}[\mu_{BB}(u) - \mu_{bb}(u)] = \sum_{u=1}^{m} c_u. \qquad (3.2.28)$$

If $F_1$ backcrosses with an individual with $BB$ type in $F_0$, then their progeny should have chance of $1/2$ and $1/2$ of all loci being $BB$ or $Bb$. Thus the contribution to the mean by the $u$th gene should be

$$\mu_u = \xi_u c_u + \frac{1}{2}(1 - \xi_u)(1 + d_u)c_u + \mu_{bb}(u), \qquad (3.2.29)$$

where $\xi_u$ is a random variable with the probability of half to be 1 and half to be 0, and $d_u$ is the dominance effect defined somewhat differently from (3.1.2). In particular, if $\xi_u = 1$, $\mu_u = \mu_{BB}(u)$, and when $\xi_u = 0$, $\mu_u = (\mu_{BB}(u) + \mu_{bb}(u))/2 + d_u c_u/2$, which reduces to additive mean $(\mu_{BB}(u) + \mu_{bb}(u))/2$ for $Bb$ when $d_u = 0$. Adding an individual index $h$ to $\mu_u$ and $\xi_u$, the trait value for individual $h$ becomes

$$\sum_{u=1}^{m} \mu_{uh} = \sum_{u=1}^{m} \xi_{uh} c_u + \frac{1}{2}(1 - \xi_{uh})(1 + d_u)c_u + \mu_{bb} + \varepsilon_h, \qquad (3.2.30)$$

where $\varepsilon_h$ represents nongenetic variation.

Apparently $\xi_{uh}$ and $\xi_{vh}$ are not independent if loci $u$ and $v$ are linked. Let $\theta_{uv}$ be the recombination fraction between them. Then it can be shown that

$$\text{Cov}(\xi_{uh},\ \xi_{vh}) = (1 - 2\theta_{uv})/4. \tag{3.2.31}$$

Suppose there are $t$ marker loci under observation. Then the relation between marker loci and gene loci may be described by Fig. 3.2.10. Here we ignore the case when there are more than one QTL between two markers. In the backcross experiment, the markers for the backcross offsprings can also be represented as $1$ or $2$. Thus, without loss of generality, we let the marker information for individual $h$ at marker locus $i$ be

$$x_{hi} = \begin{cases} 1 & \text{if the alleles are homozygatic} \\ 0 & \text{otherwise.} \end{cases} \tag{3.2.32}$$

Suppose only one polygene is in the neighborhood of markers $i-1, i, i+1$ as shown in Fig. 3.2.10. Then it is conceivable that $\beta_i$ and $\beta_{i+1}$ in the regression equation,

$$y_h = \beta_0 + \sum_{k=1}^{t} x_{hk}\beta_k + \varepsilon_h, \quad h = 1, 2, \cdots, n,$$

should show significance and the magnitude of the significance is proportional to $c_u$. According to a property of the regression coefficients (see Appendix A.1), other polygenes should not contribute to the significance of $\beta_i$ and $\beta_{i+1}$. Using (3.2.29) to (3.2.31), one can show (Zeng, 1993) that

$$\beta_i = \frac{\theta_{u(i+1)}(1 - \theta_{u(i+1)})(1 - 2\theta_{iu})}{\theta_{i(i+1)}(1 - \theta_{i(i+1)})}[(1 - d_u)c_u/2]. \tag{3.2.33}$$

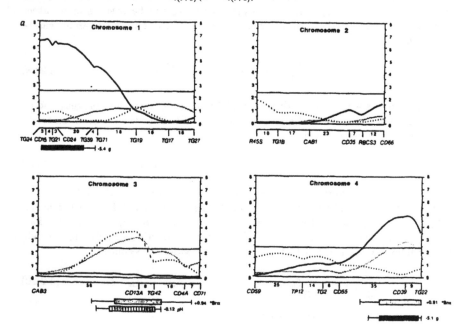

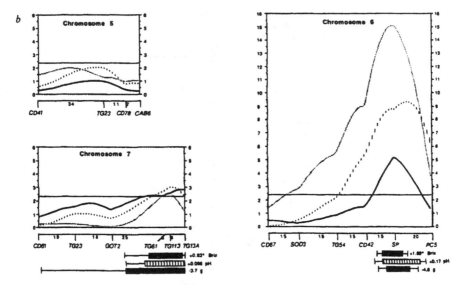

Figure 3.2.8: QTL likelihood maps on 7 out of the 12 chromosomes of tomato for fruit mass (solid lines and bars), soluble-solids concentration (dotted lines and bars) and pH (hatched lines and bars), with 70 genetic markers on the 12 chromosomes (862 cM) with an average distance 14.3 cM between markers. The RFLP map used in the analysis is presented along the abscissa in cM. The program MAPMAKER-QTL was used. The y-axis values are the LOD scores in the likelihood ratio test. Threshold 2.4 is used to claim significance with false-alarm rate less than 0.05. Bars below each graph indicate $10 : 1$ support intervals (approximate $\text{LOD}_{\max} - 1$ in Fig. 2.1.6), whereas the lines extended out from the bars indicate 100:1 support intervals. (From Paterson et al. (1988), copyright © by Macmillan Magazines Limited and the authors. Reprinted by permission.)

Indeed, $\beta_i$ depends only on gene $u$. Based on this fact, Zeng (1994) proposed the following QTL detection scheme. For any given marker $i$, let the regression equation be

$$y_h = \beta_0 + \beta^* x_h^* + \sum_{k \neq i, i+1} x_{hk}\beta_k + \varepsilon_h, \qquad (3.2.34)$$

where $\varepsilon_h \sim N(0, \sigma^2)$ with $\sigma^2$ unknown, $x_{hk}$ is defined by (3.2.32), and $x_h^*$ represents the status of a possible polygene between marker loci $i$ and $i + 1$, i.e., $x_h^* = 1$ for $BB$ and $x_h^* = 0$ for $Bb$. The value $x_h^*$ is a random variable and can be estimated by marker information at $i$ and $i + 1$. The simplest way is to use Table 3.2.1 to find $\hat{x}_h^* \equiv E[x_h^* \mid$ markers at loci $i$ and $i + 1]$. Note that $\theta_1$ in Table 3.2.1 is the $\theta_{iu}$ in Fig. 3.2.10 and $\theta_{iu}$ is assumed to be "known" in the sliding searching scheme (see Fig. 3.2.2). In this method, (3.2.34) is a standard regression model and $\beta^*$ and its statistical significance can be computed. One problem with this approach, as we have mentioned in model (3.2.15), is that $\varepsilon_h$, containing the deviation between $\hat{x}_h^*$ and $x_h^*$, may not be normally distributed. Haley and Knott (1992), Martinez and Curnow (1992) and Xu (1995)

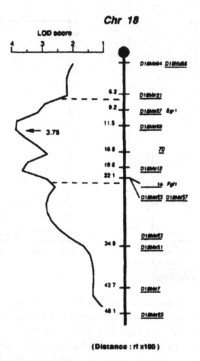

**(Distance : rf x100 )**

Figure 3.2.9: QTL likelihood maps for mouse chromosome 18 with 12 genetic markers. The markers and their distances in cM are given on the vertical line. Sixty-seven intercross progeny were used to determine the QTL for testicular weight. The LOD scores were estimated using MAPMAKER-QTL. The horizontal stippled lines adjacent to the likelihood curve define the 95% confidence interval. (From Asada et al. (1994), copyright © by Nature Genetics and the authors. Reprinted by permission.)

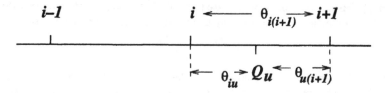

Figure 3.2.10: The marker loci are $i-1$, $i$ and $i+1$, and the $u$th polygene is at $Q_u$.

checked this approach in the original Lander and Botstein model, where $\xi_i$ in (3.2.1) was replaced by $E[\xi_i|\text{marker information}]$ and a simple linear regression was used. Little practical difference was observed in estimating the regression coefficient, although the statistical significance level by the regression method is less reliable than that by the likelihood method based on (3.2.2).

To take the likelihood approach, let the density function for $y_h$ in (3.2.34) be

$$f(y_h) = p_h(0)f_h(0) + p_k(1)f_h(1),$$ (3.2.35)

where $p_h(j) = Pr\{x_h^* = j | \text{marker and phenotype information}\}$ and

$$f_h(j) = \phi\left(\beta_0 + \beta^* j + \sum_{k \neq i, i+1} x_{hk}\beta_k, \sigma^2\right),$$

$j = 0, 1$, and $\phi(\mu, \sigma^2)$ is the normal density function for $y_h$ with mean $\mu$ and variance $\sigma^2$. Consequently, the likelihood function for $n$ progeny is

$$L(\boldsymbol{\beta}, \sigma^2 | \boldsymbol{x}) = \prod_{h=1}^{n} f(y_h),$$ (3.2.36)

where $\boldsymbol{\beta}$ denotes the parameters $\beta_0$, $\beta^*$ and $\beta_k$s and $\boldsymbol{x}$ denotes all the marker information. To find the mle of the unknown parameters $\beta$s, $\sigma^2$ is now not easy, because the $p_h(i)$s in (3.2.35) are not easy to estimate. Zeng (1994) suggested the expected/ condition maximization (ECM) algorithm by Meng and Rubin (1993). EMC is an iteration process in which $p_k(i)$ is estimated not only by the marker information but also by the value $y_h$, i.e., the posterior distribution at step $s + 1$ depends on the prior distribution at step $s$ with

$$p_h^{s+1}(i) = \frac{p_h^s(i)f_h^s(i)}{\displaystyle\sum_{i=0}^{1} p_h^s(i)f_h^s(i)}, \quad i = 0, 1,$$

where $p_h^0(i)$ at step 0 depends on the marker information only. The detailed computation can be found in Zeng (1994). The criterion to claim a polygene between loci $i$ and $i + 1$ is the likelihood ratio test assuming $H_0 : \beta^* = 0$ and $H_1 : \beta^* \neq 0$ in (3.2.36), i.e., using

$$\lambda = -2\ln\frac{L(\beta^* = 0)}{L(\hat{\beta}^*)}.$$

However, since in estimating $\beta$, the $p$s in (3.2.35) have also been estimated, $\lambda$ is not simply a $\chi^2$ distribution with 1 degree of freedom. In simulation, Zeng (1994) found that $\lambda$ is between $\chi_1^2$ and $\chi_2^2$. Thresholds for significance tests can be found in that paper.

Fig. 3.2.11 shows the advantage of using model (3.2.36). The LOD scores obtained by this method (solid curves) are higher and more articulate in locating a polygene than the other two methods; long-dashed curves were from a model which uses only unlinked markers on other chromosomes in the summation of (3.2.34) and short-dashed curves were based on Lander and Botstein's algorithm (3.2.3).

Computer program QTL-Cartographer, at web site:

$$http://statgen.ncsu.edu/qtlcart/cartographer.html,$$

is available to do the polygene detection.

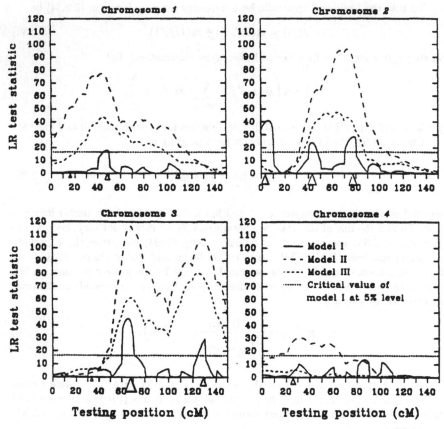

Figure 3.2.11: A simulation experiment of QTL mapping on a backcross population. The length of each chromosome is 150 cM with marker at every 10 cM. Ten QTL's were simulated with positions indicated by triangles under the abscissa. The size of the triangle is proportional to the QTL effect. The three curves are the likelihood profiles using Lander and Botstein's algorithm (short-dashed curves, Model III in the figure), unlinked markers (long-dashed curves, Model II) and (3.2.36) (solid curves, Model I). The sample size is 300 with total heritability 0.70. (From Zeng (1994), copyright ©by The Genetics Society of America. Reprinted by permission.)

## 3.3   Heritability Estimation

From previous sections, we have seen how quantitative genes may be located. However, if there is no hint where the gene is, a dense marker set has to be used in the initial search. A ten cM interval marker map means 387 markers for a human genome search (information taken from Research Genetics, Inc. www.resgen.com/Human Screening Sets). This is a tremendous investment. Moreover, the sample size depends on the heritability. If the heritability is small, it may not be worth the effort. Hence, in the beginning, we may not want to locate the gene but simply want to know the magnitude of the heritability.

In a way, heritability is difficult to measure because it interacts with the environment. A famous example is that the sex of a crocodile is determined by the ambient temperature when the egg is laid and shortly afterward. High or low temperature (typically $< 31°C$ or $> 33°C$) at early embryonic development produces females. Males are produced in the middle temperature (close to 100% males at 32°C, see Luxmoore, 1992, p. 18). However, this does not mean that the genetic factor is not important. Without the genes there is no animal and without a proper environment there is no animal either. Thus, we can talk about heritability only when the environment is *proper*. If the environment changes drastically, the heritability may be totally different.

Take intelligence as another example. There is sustained interest in knowing how much of human intelligence is inherited. Mozart's childhood was unique, but there must have been many children at his time with *similar* childhood experiences but who were unable to develop and demonstrate similar accomplishments in music, so there must be a genetic factor in it. However, if Mozart had been raised in a society without sophisticated music, he could not have developed his talent. As will be seen later in this section, approximately 50% of human intelligence and personality can be attributed to genes. But results of this kind can be controversial due to disagreement on the environmental effect and its measurement.

We have already laid down one definition of heritability in (3.1.7). Using Table 3.1.1 as an example, we have, by letting 1, 2, and 3 represent $bb$, $bB$ and $BB$ respectively,

$$\sigma_X^2 = \sum_{i=1}^{3} f_i \sigma_i^2 + \sum_{i=1}^{3} f_i (\mu_i - \bar{\mu})^2, \quad \text{where,}$$

$$\bar{\mu} = \sum_{i=1}^{3} f_i \mu_i, \quad \text{and}$$

$$\sigma_e^2 = \sum_{i=1}^{3} f_i \sigma_i^2,$$

or

$$h^2 = 1 - \frac{\sum_{i=1}^{3} f_i \sigma_i^2}{\sum_{i=1}^{3} f_i \sigma_i^2 + \sum_{i=1}^{3} f_i (\mu_i - \bar{\mu})^2}. \tag{3.3.1}$$

One thing counterintuitive in this formula is that heritability depends on the allele frequencies $f_i$. Thus, the heritability changes in the same environment when the

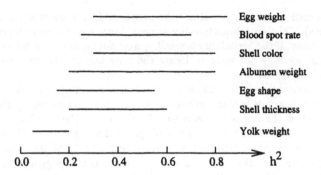

Figure 3.3.1: Range of heritability $h^2$ reported for a variety of chicken egg characters. (Redrawn from Suzuki, et al. (1981).)

population composition changes. Hence, to interpret the heritability, it is necessary to specify both the population and the environment. Fig. 3.3.1 gives the heritability of various chicken egg characteristics raised in farms. Even under strict control on most chicken farms the heritability varies considerably.

Human heritability has to be estimated from the phenotype information of relatives. We will start with twin studies.

### 3.3.1  Twin Studies

There are two types of twins; monozygotic twins (MZ), who share the same genes and dizygotic twins (DZ), whose genetic compositions are like siblings. The twins can be reared together or apart. Thus, there are four cases in twins studies:

MZT = monozygotic twins reared together,
DZT = dizygotic twins reared together,
MZA = monozygotic twins reared apart,
DZA = dizygotic twins reared apart.

Let $X_{i1}$ and $X_{i2}$ denote the attributes of the trait under investigation for the two twins from the $i$th family in a sample of size $n$. Define

$$X_{ij} = \mu + u_{ij} + e_{ij} + \varepsilon_{ij}, \quad i = 1,\ldots,n, \quad j = 1,2, \tag{3.3.2}$$

where $\mu$ is the overall mean, $u_{ij}$ is the genetic contribution, $e_{ij}$ is the environmental (familial) contribution, and $\varepsilon_{ij}$ is all the other variations. We further assume $u$, $e$ and $\varepsilon$ are all independent with 0 mean and variances $\sigma_g^2 \equiv \sigma_u^2$, $\sigma_e^2$ and $\sigma_\varepsilon^2$. Thus, $V[X_{ij}] = \sigma_g^2 + \sigma_e^2 + \sigma_\varepsilon^2$ and

$$h^2 = \sigma_g^2/(\sigma_g^2 + \sigma_e^2 + \sigma_\varepsilon^2). \tag{3.3.3}$$

In order to estimate the heritability, the following three statistics are usually used.

Table 3.3.1: Expected values for within ($W$) and between ($B$) MSE and $\rho$ defined in (3.3.4).

|  | $\rho$ | $E[W]$ | $E[B]$ |
|---|---|---|---|
| MZT | $(\sigma_g^2 + \sigma_e^2)/\sigma_X^2$ | $\sigma_\varepsilon^2$ | $2\sigma_g^2 + 2\sigma_e^2 + \sigma_\varepsilon^2$ |
| MZA | $\sigma_g^2/\sigma_X^2$ | $\sigma_e^2 + \sigma_\varepsilon^2$ | $2\sigma_g^2 + \sigma_e^2 + \sigma_\varepsilon^2$ |
| DZT | $(C\sigma_g^2 + \sigma_e^2)/\sigma_X^2$ | $(1 - C)\sigma_g^2 + \sigma_\varepsilon^2$ | $(1 + C)\sigma_g^2 + 2\sigma_e^2 + \sigma_\varepsilon^2$ |
| DZA | $C\sigma_g^2/\sigma_X^2$ | $(1 - C)\sigma_g^2 + \sigma_e^2 + \sigma_\varepsilon^2$ | $(1 + C)\sigma_g^2 + \sigma_e^2 + \sigma_\varepsilon^2$ |

1. Sample correlation $\rho \equiv \text{Cov}(X_{i1}, X_{i2})/\sigma_X^2$;

2. Within MSE (variance) $W \equiv \sum_{i=1}^{n}(X_{i1} - X_{i2})^2/(2n)$; $\qquad$ (3.3.4)

3. Between MSE (variance) $B \equiv 2\sum_{i=1}^{n}(\bar{X}_{i\cdot} - \bar{X}_{\cdot\cdot})^2/(n - 1)$;

where $\bar{X}_{i\cdot} = (X_{i1} + X_{i2})/2$ and $\bar{X}_{\cdot\cdot} = \sum_{i=1}^{n} \bar{X}_{i\cdot}/n$. Definitions in (3.3.4) are the usual definitions in bivariate data analysis and one-way analysis of variance. Table 3.3.1 gives their expected values (see Table 2 of Tellegen et al., 1988), where $C$ is the covariance between the genetic effect of two dizygotic twins (see derivations below). Only the values for DZT are derived here. For two DZT twins,

$$X_{i1} = \mu + u_{i1} + e_{i1} + \varepsilon_{i1}$$
$$X_{i2} = \mu + u_{i2} + e_{i2} + \varepsilon_{i2}.$$

So,

$$\text{Cov}(X_{i1}, X_{i2}) = \text{Cov}(u_{i1}, u_{i2}) + \sigma_e^2$$
$$\equiv C\sigma_g^2 + \sigma_e^2,$$

which is the table value. The derivation of E[W] is easy and those of E[B] follow from

$$\bar{X}_{i\cdot} = \mu + (u_{i1} + u_{i2})/2 + e_i + \bar{\varepsilon}_1, \quad V[\bar{X}_{i\cdot}] = \frac{1}{2}(1 + C)\sigma_g^2 + \sigma_e^2 + \frac{1}{2}\sigma_\varepsilon^2,$$

and the mean and variance of a $\chi^2$ distribution.

**Example 3.3.1.** *Table 3.3.2 is taken from the Minnesota twin study (Tellegen et al., 1988). Find the heritabilities of these personalities. The sample sizes are* $MZA(n = 44)$, $DZA(n = 27)$, $MZT(n = 217)$, $DZT(n = 114)$.

**Sol:** Since the $\rho$ of MZA is the same as the heritability $h^2$ and

$$\hat{\rho} = \frac{E[B] - E[W]}{E[B] + E[W]}, \qquad (3.3.5)$$

Table 3.3.2: Between (B) and Within (W) MSE of twins reared apart and together for personality trait by questionnaires.

| Scale | MZA | | DZA | | MZT | | DZT | |
|---|---|---|---|---|---|---|---|---|
| | B | W | B | W | B | W | B | W |
| Well-Being | 1.62 | 0.56 | 1.85 | 1.28 | 1.52 | 0.40 | 1.13 | 0.71 |
| Social Potency | 1.90 | 0.53 | 1.10 | 0.64 | 1.67 | 0.35 | 1.00 | 0.86 |
| Achievement | 1.33 | 0.61 | 1.25 | 1.09 | 1.58 | 0.52 | 1.02 | 0.78 |
| Social Closeness | 1.32 | 0.71 | 1.53 | 0.83 | 1.46 | 0.40 | 1.37 | 0.84 |
| Stress Reaction | 1.95 | 0.46 | 1.03 | 0.59 | 1.50 | 0.47 | 1.16 | 0.72 |
| Alienation | 1.21 | 0.42 | 1.36 | 0.94 | 1.58 | 0.46 | 1.36 | 0.61 |
| Aggression | 1.09 | 0.40 | 1.13 | 1.00 | 1.49 | 0.59 | 1.13 | 0.85 |
| Control | 1.87 | 0.61 | 1.08 | 1.02 | 1.32 | 0.55 | 0.94 | 1.06 |
| Harm Avoidance | 1.25 | 0.42 | 0.79 | 0.49 | 1.63 | 0.47 | 1.25 | 0.89 |
| Traditionalism | 1.66 | 0.51 | 1.78 | 0.78 | 1.39 | 0.46 | 1.55 | 0.56 |
| Absorption | 1.90 | 0.46 | 1.25 | 0.82 | 1.50 | 0.51 | 1.34 | 0.57 |
| Positive Emotionality | 1.29 | 0.62 | 0.95 | 1.09 | 1.70 | 0.38 | 1.11 | 0.77 |
| Negative Emotionality | 1.58 | 0.37 | 1.44 | 0.80 | 1.50 | 0.45 | 1.39 | 0.58 |
| Constraint | 1.43 | 0.39 | 0.82 | 0.76 | 1.63 | 0.44 | 1.28 | 0.77 |

*Note*: MZA = Monozygotic twins reared apart, DZA = dizygotic twins reared apart, MZT = monozygotic twins reared together, and DZT = dizygotic twins reared together.
*Source*: Tellegen et al. (1988) Copyright © 1988 by the American Psychological Association. Reprinted with permission.

$h^2$ can be estimated from Table 3.3.2. For example, the heritability of wellbeing is

$$\hat{\rho} = \frac{1.62 - 0.56}{1.62 + 0.56} = 0.48.$$

The other heritabilities are given in Table 3.3.3. The MZA column shows that the heritability of personality is approximately 50%. However, it is amazing that the correlation between monozygotic twins reared apart is in general higher than is that of the dizygotics reared together. Though the sample sizes of MZT and DZT are larger, they alone cannot be used to estimate heritability. The reason is that $\sigma_g^2$ and $\sigma_e^2$ are confounded in both cases. □

Hypothesis testing and confidence intervals can be obtained using Fisher's log transformation (see Appendix A.1)

$$z \equiv \frac{1}{2}\ln\frac{1+\hat{\rho}}{1-\hat{\rho}} \sim N(0, 1/(n-3)). \qquad (3.3.6)$$

**Example 3.3.2**  *Is there enough statistical evidence to say that the correlation on an achievement scale for MZA is higher than for DZT?*

Table 3.3.3: Sample correlations $\rho$ for personality trait by questionnaires.

| Scale | MZA | DZA | MZT | DZT |
|-------|-----|-----|-----|-----|
| Sample size | 44 | 27 | 217 | 114 |
| Primary | | | | |
| Wellbeing | .48 | .18 | .58 | .23 |
| Social potency | .56 | .27 | .65 | .08 |
| Achievement | .36 | .07 | .51 | .13 |
| Social closeness | .29 | .30 | .57 | .24 |
| Stress reaction | .61 | .27 | .52 | .24 |
| Alienation | .48 | .18 | .55 | .38 |
| Aggression | .46 | .06 | .43 | .14 |
| Control | .50 | .03 | .41 | −.06 |
| Harm avoidance | .49 | .24 | .55 | .17 |
| Traditionalism | .53 | .39 | .50 | .47 |
| Absorption | .61 | .21 | .49 | .41 |
| Higher order | | | | |
| Positive emotionality | .34 | −.07 | .63 | .18 |
| Negative emotionality | .61 | .29 | .54 | .41 |
| Constraint | .57 | .04 | .58 | .25 |

*Source:* Tellegen et al. (1988) Copyright © 1988 by the American Psychological Association. Reprinted with permission.

**Sol:** Let

$$\Delta Z = \frac{1}{2}\ln\frac{1+\hat{\rho}_1}{1-\hat{\rho}_1} - \frac{1}{2}\ln\frac{1+\hat{\rho}_2}{1-\hat{\rho}_2}$$
$$= \frac{1}{2}\ln\frac{1+0.36}{1-0.36} - \frac{1}{2}\ln\frac{1+0.13}{1-0.13} = 0.245.$$

Thus, the $z$-score for the test using (3.3.6) is

$$z = 0.245/\sqrt{1/44 + 1/114} = 1.34.$$

This is not significant at the 0.05 level. Though the differences in most single items do not show statistical significance, the combined effect is definitely significant. The probability that all the 13 correlations are higher for MZA by chance has $p$-value $2^{-12} = 0.0002$ by a two-sided sign test. □

To use all of the information for the variance components, the maximum likelihood method may be used with normality assumption. Let $(X_{i1}, X_{i2})$ have a bivariate normal distribution; the correlations in terms of the unknown parameters are given in Table 3.3.1. The likelihood equation is

$$L = \prod_{i=1}^{4}\prod_{j=1}^{n_i} \frac{1}{2\pi|\Sigma_i|^{1/2}} e^{-\frac{1}{2}(x_{ij}-\mu)'\Sigma_i^{-1}(x_{ij}-\mu)}, \tag{3.3.7}$$

Table 3.3.4: Estimates of genetic and familial inference in proportion
(all data combined).

| Scale | Variance component | | | |
|---|---|---|---|---|
| | Genetic | C parameter | Shared (familial) | Unshared |
| Wellbeing | $.48^a$ (.08) | .29 (.16) | .13 (.09) | $.40^a$ (.04) |
| Social potency | $.54^a$ (.07) | $.05^a$ (.21) | .10 (.08) | $.36^a$ (.04) |
| Achievement | $.39^a$ (.10) | .13 (.27) | .11 (.11) | $.51^a$ (.05) |
| Social closeness | $.40^a$ (.08) | .19 (.22) | $.19^a$ (.09) | $.41^a$ (.05) |
| Stress reaction | $.53^a$ (.04) | .49 (.17) | $.00^b$ | $.47^a$ (.04) |
| Alienation | $.45^a$ (.13) | $.50^b$ | .11 (.12) | $.44^a$ (.04) |
| Aggression | $.44^a$ (.05) | .27 (.19) | $.00^b$ | $.56^a$ (.05) |
| Control | $.44^a$ (.05) | $.00^{a,b}$ | $.00^b$ | $.56^a$ (.05) |
| Harm avoidance | $.55^a$ (.04) | .31 (.15) | $.00^b$ | $.45^a$ (.04) |
| Traditionalism | $.45^a$ (.10) | $.50^b$ | .12 (.10) | $.43^a$ (.04) |
| Absorption | $.50^a$ (.10) | $.50^b$ | .03 (.10) | $.47^a$ (.04) |
| Positive emotionality | $.40^a$ (.08) | $.00^{a,b}$ | $.22^a$ (.07) | $.38^a$ (.04) |
| Negative emotionality | $.55^a$ (.11) | $.50^b$ | .02 (.11) | $.43^a$ (.04) |
| Constraint | $.58^a$ (.04) | .40 (.14) | $.00^b$ | $.43^a$ (.04) |

Note: Standard errors are in parentheses.
[a] Significantly different from null value at $p < .05$.
[b] Boundary solution; therefore, no standard error computed.
Source: Tellegen et al. (1988). Copyright ©1988 by the American Psychological Association.
Reprinted with permission.

where the data are $x_{ij} = \begin{pmatrix} x_{ij1} \\ x_{ij2} \end{pmatrix}$, $i = 1, 2, 3, 4; j = 1, 2, \ldots, n_i$, and $\Sigma_i$ is the correlation matrix for case $i = $ MZT, MZA, DZT and DZA. There is no analytic solution for all the parameters. A numerical solution has to be applied. The accuracy of the solution can be estimated by evaluating the Fisher's information matrix

$$\vartheta = E\left[\frac{\partial^2 \ln L}{\partial \theta_i \partial \theta_j}\right],$$

where $\theta_i$ and $\theta_j$ are two of the parameters. Heritability by MLE of the twin study in Example 3.3.1 with all the four cases combined in the estimation process is given in Table 3.3.4. We see that the genetic contribution to main human characters are nearly 50%. A recent study on IQ heritability based on meta-analysis of many reports can be found in Devlin et al. (1997).

## 3.3.2  Extension to Simple Pedigrees

To extend the twin results beyond siblings, we need an indicator to quantify the genetic relationship between two relatives. It is defined as the **kinship coefficient** $\phi(A, B)$ between two individuals $A$ and $B$ by

$\phi(A, B) = P\{$At any given locus, two randomly picked alleles, one from A and one from B are IBD$\}$

For example, $\phi(A, B) = 1/2$ for monozygotic twins $A$ and $B$, if their parents are unrelated; and $= 1/4$ for two siblings, if their parents are unrelated. This coefficient is very important in quantitative trait study. It will be referred to many times in the future. To see how this coefficient is useful, let the trait value due to genes of individual $i$ be

$$u_i = a_i^m + a_i^p + d_i, \qquad (3.3.8)$$

where $a_i^m$ and $a_i^p$ are the genetic values determined from the person's maternal and paternal side, respectively, in the additive sense and $d_i$ is the dominance effect. Moreover, we may assume that $a_i^m + a_i^p$ and $d_i$ are uncorrelated. To see how this can be assumed, we need to examine more closely on the decomposition (3.3.8).

For given trait $u_i$, the partitioning of $u_i = a_i^m + a_i^p + d_i$ is not unique. Usually, the priority is given to the additive effect, i.e., we consider dominance effect only when the additive effects cannot explain the variation in $u_i$. We use Table 3.1.1, (3.1.2) and HWE frequencies $f_{bb} = q^2, f_{bB} = 2pq$ and $f_{BB} = q^2$ as an example. Let the additive effect for allele $B$ be $\alpha_B$ and $b$ be $\alpha_b$. Thus the additive effects for $BB$, $Bb$ and $bb$ are, respectively, $2\alpha_B, \alpha_B + \alpha_b,$ and $2\alpha_b$. Let $\delta_{BB}, \delta_{Bb},$ and $\delta_{bb}$ be the dominant effects for $BB$, $Bb$ and $bb$ respectively. Then

$$\gamma_{BB} = 2\alpha_B + \delta_{BB} = a,$$
$$\gamma_{Bb} = \alpha_B + \alpha_b + \delta_{Bb} = d,$$
$$\gamma_{bb} = 2\alpha_b + \delta_{bb} = -a.$$

Suppose the parents of individual $i$ are unrelated. Then both $a_i^m$ and $a_i^p$ have probability $p$ to be $\alpha_B$ and $q$ to be $\alpha_b$, and $d_i$ has probabilities $p^2, 2pq$, and $q^2$ to be $\delta_{BB}, \delta_{Bb}$ and $\delta_{bb}$ respectively. Since we can always put the nonzero part of mean into the overall mean $\mu$ in (3.1.1), without loss of generality, we let

$$E(u_i) = E(a_i^m) = E(a_i^p) = E(d_i) = 0.$$

Note that the subscript $i$ is the individual index. Since we want the variance component due to dominance effect to be as small as possible, we should find $\alpha_B$ and $\alpha_b$ such that

$$\text{Var}(d_i) = p^2(a - 2\alpha_B)^2 + 2pq(d - \alpha_B - \alpha_b)^2 + q^2(-a - 2\alpha_b)^2$$

is as small as possible. By setting the partial derivative of $\text{Var}(d_i)$ with respect to $\alpha_B$ to 0, i.e.,

$$0 = (a - 2\alpha_B)p + q(d - \alpha_B - \alpha_b) = \delta_{BB}p + \delta_{Bb}q$$

and using $E(a_i^m) = \alpha_B p + \alpha_b q = 0$, we have

$$\alpha_B = ap + dq.$$

Similarly, we have

$$\delta_{Bb}p + \delta_{bb}q = 0, \quad \text{and} \quad \alpha_b = dp - aq.$$

Moreover,

$$\text{Cov}(a_i^m + a_i^p, d_i) = E[d_i(a_i^m + a_i^p)]$$
$$= p^2 \delta_{BB}(2\alpha_B) + 2pq\delta_{Bb}(\alpha_B + \alpha_b) + q^2\delta_{bb}(2\alpha_b) = 0,$$

or $(a_i^m + a_i^p)$ and $d_i$ are uncorrelated, and the variances agree with the two variance components given in (3.1.5),

$$\text{Var}(a_i^m + a_i^p) = 2pq[a - d(p - q)]^2 \quad \text{and} \quad \text{Var}(\delta_i) = 4p^2q^2d^2. \tag{3.3.9}$$

For more than two allele models, the results are similar. Details can be found in Lange (1997, pp. 87–89).

Returning to (3.3.8), we define the trait value for person $j$ as

$$u_j = a_j^m + a_j^p + d_j.$$

For a person with unrelated parents,

$$V[u_j] = V[a_j^m] + V[a_j^p] + V[d_j] \equiv \sigma_m^2 + \sigma_p^2 + \sigma_d^2 \equiv \sigma_a^2 + \sigma_d^2. \tag{3.3.10}$$

The last equality implies $\sigma_m^2 = \sigma_p^2 = \sigma_a^2/2$. By an elementary statistical formula and the independence of $a$'s and $d$, we have

$$\text{Cov}(u_i, u_j) = \text{Cov}(a_i^m, a_j^m) + \text{Cov}(a_i^m, a_j^p) + \text{Cov}(a_i^p, a_j^m) + \text{Cov}(a_i^p, a_j^p) + \text{Cov}(d_i, d_j). \tag{3.3.11}$$

The following lemma is useful to relate the kinship coefficient to (3.3.11).

**Lemma 3.3.1** *Suppose that two subjects A and B will be picked in a population and that there is probability p that A and B are the same subject and probability $1 - p$ that A and B are two randomly picked subjects. Then $\text{Cov}(u_A, u_B) = pV(u_A) = pV(u_B)$, where $u_A$ and $u_B$ are the values of A and B.*

**Remark** Two randomly picked individuals may have the same value $X_A = X_B$, but their covariance can still be 0. This is similar to the situation where the covariance between two independently tossed dice is 0, but they may have the same value when tossed.

Since two alleles, if not IBD, are equivalent to being picked at random, we have

$$\text{Cov}(a_i^m, a_j^m) = \sigma_m^2 P\{a_i^m = a_j^m \text{ by IBD}\} = \sigma_a^2/2 \cdot P\{a_i^m = a_j^m \text{ by IBD}\} \tag{3.3.12}$$

The above equation is also true with the exchange of $m$ and $p$ for all the first four terms in (3.3.11) which becomes

$$\text{Cov}(u_i, u_j) = 2\sigma_a^2\phi(u_i, u_j) + \text{Cov}(d_i, d_j), \tag{3.3.13}$$

where $\phi(u_i, u_j)$ is the kinship coefficient between individuals $i$ and $j$. Since the dominance effect happens only when the two alleles are heterogeneous, we have

$$\text{Cov}(d_i, d_j) = \sigma_d^2 P\{\text{Both i and j have the same dominance effect by IBD}\}$$
$$\equiv \sigma_d^2 \Delta_7(i, j), \tag{3.3.14}$$

Table 3.3.5: Kinship coefficient $\phi$ and Jackquard's $\Delta_7$.

|  | $\phi$ | $\Delta_7$ |
|---|---|---|
| Identical twins | 1/2 | 1 |
| Parent-offspring | 1/4 | 0 |
| Full sibs | 1/4 | 1/4 |
| Full cousins | 1/16 | 0 |

where $\Delta_7(i,j)$ is called **Jackquard's $\Delta_7$ coefficient** between individuals $i$ and $j$ (see Jackquard, 1974, pp. 106–7). It is the same as

$$\Delta_7(i,j) = P\{g_i^m = g_j^p \neq g_i^p = g_j^m \text{ or } g_i^m = g_j^m \neq g_i^p = g_j^p \text{ by IBD}\},$$

where $g_i^m$, $g_i^p$, and so on denote the genes by IBD to individual $i$ from the mother's and father's sides, etc. Table 3.3.5 gives the kinship and $\Delta_7$ coefficients for four close relatives when their parents are not related.

**Example 3.3.3** *Find the correlation, in terms of $\sigma_a^2$, $\sigma_d^2$, $\sigma_e^2$ and $\sigma_\varepsilon^2$, between two siblings raised under the same environment. (Assume that their parents are not related.)*

**Sol:** By (3.3.2), (3.3.13), (3.3.14) and Table 3.3.5, we have

$$\text{Cov}(X_1, X_2) = \sigma_a^2/2 + \sigma_d^2/4 + \sigma_e^2.$$

Thus, the correlation becomes

$$\rho = \frac{\sigma_a^2/2 + \sigma_d^2/4 + \sigma_e^2}{\sigma_a^2 + \sigma_d^2 + \sigma_e^2 + \sigma_\varepsilon^2}. \qquad \square$$

Note that $\sigma_g^2$ in (3.3.3) is the same as $\sigma_a^2 + \sigma_d^2$. Thus, using DZT in Table 3.3.1, the coefficient $C$ becomes

$$C = \frac{\sigma_a^2 + \sigma_d^2/2}{2(\sigma_a^2 + \sigma_d^2)}.$$

To have some insight into the two variance components $\sigma_a^2$ and $\sigma_d^2$, let us examine the situation in Table 3.3.6.

In this table, we see that the phenotype of an offspring is determined totally by the interaction between the two alleles. Thus, we cannot say a strong father or mother will produce a strong child, assuming "stronger" means a larger trait value. Even if both mother and father are strong, this will not guarantee a stronger child. On the contrary,

Table 3.3.6: An example of $\sigma_a^2 = 0$ and $\sigma_d^2 \neq 0$.

| Genotype | bb | Bb | BB |
|---|---|---|---|
| Population frequency | 0.25 | 0.50 | 0.25 |
| Trait value | 10 | 15 | 10 |

only two weak parents with different genotypes can guarantee a strong child. This happens for many trait of domestic plants. The seeds produced by the high-yields plants cannot be used for the next generation. Good seeds have to be purchased from a seed company every year. Some geneticist do not consider the dominance effect *inheritable*. They would rather use another definition of heritability called **heritability in the narrow sense**

$$h_n^2 = \frac{\sigma_a^2}{\sigma_a^2 + \sigma_d^2 + \sigma_e^2 + \sigma_\varepsilon^2}. \tag{3.3.15}$$

Heritability based on the new definition can be derived using the same techniques.

We have seen that estimating the heritability is equivalent to estimating the variance components. Suppose we have pedigrees from many families. Let $X = (x_1, x_2, \ldots, x_k)'$ denote the vector of the $k$ members of a pedigree. With normality assumption, we let

$$X \sim N(\mu, \Sigma), \quad \text{with} \quad \Sigma = 2\sigma_a^2 \Sigma_a + \sigma_d^2 \Sigma_d + \sigma_e^2 \Sigma_e + \sigma_\varepsilon^2 I, \tag{3.3.16}$$

where $\mu$ is a constant vector, $I$ is the identity matrix, and all the $\Sigma$s denote the relations between the family members. More precisely,

$\Sigma_a$ = the kinship coefficients between the members

$\Sigma_d$ = the $\Delta_7$ coefficients between the members

$\Sigma_e$ = the shared environmental status between the members

The likelihood equation is almost identical to (3.3.7) in all aspects except number of parameters becomes larger. If (A.2.4) is used to compute power in hypothesis testing, the information matrix can also be constructed by

$$E\left(-\frac{\partial^2 L}{\partial \mu^2}\right) = 1'\sigma^{-1}1,$$

$$E\left(-\frac{\partial^2 L}{\partial \mu \partial \sigma_i}\right) = 0,$$

$$E\left(-\frac{\partial^2 L}{\partial \sigma_i^2 \partial \sigma_j^2}\right) = \frac{1}{2} tr\left(\Sigma^{-1} \frac{\partial \Sigma}{\partial \sigma_i^2} \Sigma^{-1} \frac{\partial \Sigma}{\partial \sigma_j^2}\right),$$

where $1$ is vector of 1s. Examples can be found in Schork and Schork (1993) where simple pedigrees such as two parents with several children were considered.

### 3.3.3 Computation of kinship coefficients

There are two basic facts to keep in mind when computing kinship coefficients.

**Lemma 3.3.2** *If the two parents $F$ and $M$ of two identical twins $A$ and $B$ have a kinship coefficient $\phi(F, M)$, then the kinship coefficient between $A$ and $B$ is*

$$\phi(A, B) = (1 + \phi(F, M))/2.$$

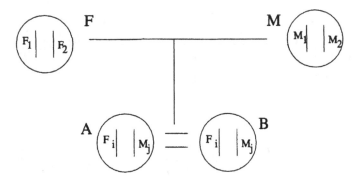

Figure 3.3.2: Illustration for Lemma 3.3.2. $F$ and $M$ stand for two parents and $A$ and $B$ represent two identical twins.

PROOF. In Fig. 3.3.2, there are four equal possibilities to choose $(F_i, F_i)$, $(F_i, M_j)$, $(M_j, F_i)$, or $(M_j, M_j)$ from $A$ and $B$. In other words,

$$\phi(A, B) = \frac{1}{4}[P(F_i = F_i \text{ by IBD}) + P(F_i = M_i \text{ by IBD}) + P(M_j = F_i \text{ by IBD})$$
$$+ [P(M_j = M_j \text{ by IBD})]$$
$$= \frac{1}{4}[1 + \phi(F, M) + \phi(F, M) + 1] = (1 + \phi(F, M))/2. \qquad \square$$

**Corollary** For any individual A with father F and mother M, his kinship coefficient to him/her-self is $\phi(A, A) = (1 + \phi(F, M))/2$. This $\phi(F, M)$ is called the **inbreeding coefficient** of individual $A$. It is an important indicator in population genetics (e.g., Roff, 1997).

**Lemma 3.3.3** (*Basic formula in kinship computation*). *Suppose A and B are two distinct individuals and the parents of A are FA and MA. Then,*

$$\phi(A, B) = (\phi(FA, B) + \phi(MA, B))/2. \qquad (3.3.17)$$

PROOF. Using Fig. 3.3.3,

$$\phi(A, B) = P\{\text{An allele from A} = \text{An allele from B by IBD}\}$$
$$= P\{F_i = \text{An allele from B by IBD}\}P\{F_i \text{ is chosen}\}$$
$$+ P\{M_j = \text{An allele from B by IBD}\}P\{M_j \text{ is chosen}\}$$
$$= (\phi(FA, B) + \phi(MA, B))/2.$$

**Example 3.3.4** *Compute the kinship coefficient matrix for the seven individuals in Fig. 3.3.4.*

**Sol:** Let matrix $\Phi_k = (\phi_{ij})$ where $\phi_{ij}$ denotes the kinship coefficient between individuals $i$ and $j$ for $k$ individuals.

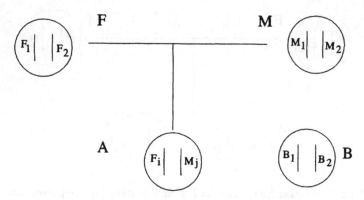

Figure 3.3.3: Illustration for Lemma 3.3.3. $F$, $M$ and $A$ stand for two parents and a child, while $B$ represents an outside family member.

The rules for iterative computation as $k$ increases are:

1. Start with the highest level of the pedigree. Any two persons with unknown ancestors are assumed to be unrelated, i.e., $\phi(1,2) = \phi(1,5) = \phi(2,5) = 0$ in Fig. 3.3.4.

2. At step $k$, let

$$\Phi_k = \begin{bmatrix} \Phi_{k-1} & \Phi_{k-1} s_k \\ s_k' \Phi_{k-1} & \phi_{kk} \end{bmatrix}, \quad s_k = \begin{bmatrix} s_{1k} \\ s_{2k} \\ \vdots \\ s_{k-1,k} \end{bmatrix},$$

where $s_{ji} = \frac{1}{2}$ if $j$ is the father or mother of $i$, and $s_{ji} = 0$ otherwise, based on Lemma 3.3.3.

3. The value of $\phi_{kk}$ can be computed using the corollary of Lemma 3.3.2.

Go back to Fig. 3.3.4. To start, we have

$$\Phi_2 = \begin{bmatrix} 1/2 & 0 \\ 0 & 1/2 \end{bmatrix}.$$

For $\Phi_3$,

$$\Phi_{3-1} s_3 = \begin{bmatrix} 1/2 & 0 \\ 0 & 1/2 \end{bmatrix} \begin{bmatrix} 1/2 \\ 1/2 \end{bmatrix},$$

and $\phi_{33} = (1 + \phi_{12})/2 = 1/2$, or

$$\Phi_3 = \begin{bmatrix} 1/2 & 0 & 1/4 \\ 0 & 1/2 & 1/4 \\ 1/4 & 1/4 & 1/2 \end{bmatrix}.$$

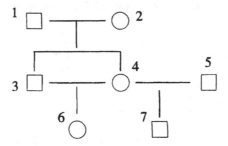

Figure 3.3.4: Illustration for kinship coefficient computation for Example 3.3.4.

For $\Phi_4$,

$$\Phi_3 s_3 = \begin{bmatrix} 1/2 & 0 & 1/4 \\ 0 & 1/2 & 1/4 \\ 1/4 & 1/4 & 1/2 \end{bmatrix} \begin{bmatrix} 1/2 \\ 1/2 \\ 0 \end{bmatrix},$$

and $\phi_{44} = (1 + \phi_{12})/2 = 1/2$, or

$$\Phi_4 = \begin{bmatrix} 1/2 & 0 & 1/4 & 1/4 \\ 0 & 1/2 & 1/4 & 1/4 \\ 1/4 & 1/4 & 1/2 & 1/4 \\ 1/4 & 1/4 & 1/4 & 1/2 \end{bmatrix}.$$

After $\Phi_5$, we have $\Phi_6$,

$$\Phi_6 = \begin{bmatrix} 1/2 & 0 & 1/4 & 1/4 & 0 & 1/4 \\ 0 & 1/2 & 1/4 & 1/4 & 0 & 1/4 \\ 1/4 & 1/4 & 1/2 & 1/4 & 0 & 3/8 \\ 1/4 & 1/4 & 1/4 & 1/2 & 0 & 3/8 \\ 0 & 0 & 0 & 0 & 1/2 & 0 \\ 1/4 & 1/4 & 3/8 & 3/8 & 0 & 5/8 \end{bmatrix}.$$

Note that $\phi_{66} = (1 + 1/4)/2 = 5/8$. Finally, we have

$$\Phi_7 = \begin{bmatrix} 1/2 & 0 & 1/4 & 1/4 & 0 & 1/4 & 1/8 \\ 0 & 1/2 & 1/4 & 1/4 & 0 & 1/4 & 1/8 \\ 1/4 & 1/4 & 1/2 & 1/4 & 0 & 3/8 & 1/8 \\ 1/4 & 1/4 & 1/4 & 1/2 & 0 & 3/8 & 1/4 \\ 0 & 0 & 0 & 0 & 1/2 & 0 & 1/4 \\ 1/4 & 1/4 & 3/8 & 3/8 & 0 & 5/8 & 3/16 \\ 1/8 & 1/8 & 1/8 & 1/4 & 1/4 & 3/16 & 1/2 \end{bmatrix}.$$

The SAS INBREED procedure can be used to compute the $D = \Phi$. The program for this example is given in Appendix B.7.

Table 3.3.5 can now be extended to more relatives in Table 3.3.7.

Table 3.3.7: (Extension of Table 3.3.5) Kinship $\phi$ and $\Delta_7$.

|                        | $\phi$ | $\Delta_7$ |
|------------------------|--------|------------|
| Grand parent–child     | 1/8    | 0          |
| Half-sibs              | 1/8    | 0          |
| Uncle-nephew           | 1/8    | 0          |
| Offspring of sib-mating| 3/8    | 7/3        |

The kinship coefficient for two persons can be generalized to an order list of $n$ people. Detail of this generalization can be found in Lange (1997).

## 3.4    Breeding Value Evaluation Models

Using artificial mating to produce more desirable crops and animals has been practiced consciously and unconsciously for thousands of years with dramatic results. Though gene identification will certainly add a new dimension to this practice, it may not be able to completely replace the traditional methods. Lande and Thompson (1990) stated the following four reasons (sources of evidence are cited in that paper):

1. The rate of improvement of economically important characters, such as grain yield in corn and wheat or milk yield in dairy cattle, has been a few to several per cent of the mean per year for the past several decades. For various crop plants it has been established that roughly half of their progress is due to improvement in husbandry practices. In other words, improvement of genetics alone will not give big leaps due to the environmental and genetic interaction. Genetic change has to wait until the environment is right.

2. Most characters of economic importance are quantitative traits, influenced by numerous loci throughout the genome that often have individually small effects. Genes with small effect are difficult to map precisely, and there may be practical problems in engineering polygenic traits once the genes have been identified at the molecular level.

3. Single genes of major effect that are amenable to genetic engineering usually have a deleterious side effect, called **pleiotropic**. This helps to explain why natural evolution usually proceeds by a series of small genetic steps. Genetic engineers recently have confronted this fact in the form of low viability and fertility of mice and pigs expressing transgenic growth hormones. They may need to be amended by further selection outside the laboratory.

4. The high mutability of polygenic characters guarantees genetic variation will arise naturally in populations. These variations can be usefully selected to improve on whatever previous gains have been made. For a typical quantitative character, in excess of one per hundred gametes contains a new mutation having a detectable effect. New additive genetic variance arises by mutation at a rate on the order of $10^{-3}$ times

the environmental variance per generation in characters in variety of species. At these rates, spontaneous mutation has been judged to be important in long-term selection programs lasting more than 20 generations.

One method of combining molecular genetics and traditional artificial selection is **marker-assisted selection (MAS)**. We will start with the classic selection model.

## 3.4.1 Henderson Model

Breeding-value models are used to evaluate the breeding value of animals or plants. Let there be $i = 1, 2, \ldots, n$ individuals. The classic mixed models are usually used to represent the relation between the phenotype $y_i$ of individual $i$ and 1) genetic related effects $u_j$, $j = 1, \ldots, r$ and 2) environment effects $\beta_j, j = 1, \ldots, k$. More specifically,

$$y_i = \sum_{j=1}^{k} x_{ij}\beta_j + \sum_{j=1}^{r} z_{ij}u_j + e_i, \quad i = 1, 2, \ldots, n, \qquad (3.4.1)$$

where $x_{ij}$ and $z_{ij}$ specify the relation between $y_i$ and the $\beta_j$s and $u_i$s, and $e_i$ is the error term. For example, when $y_i = $ milk production of the $i$th cow, we may have $x_{i1} = 1$, $x_{i2} = diet$, her father was sire $u_3$, and her mother was dam $u_8$. Thus, $\beta_1$ is the overall mean, $\beta_2$ is the diet effect, and all the $z_{ij}$'s are 0 except $j = 3$ and 8. In animal breeding, the $u$s are considered random effects. The reason is, the effects of father $u_j$ to his daughters is a random variable subject to genetic variation in meiosis. However, when the data are given, we can also evaluate the $u_j$ for any particular individual. To see the logic behind this, we can use simple linear regression as an example. In the model $y_i = \alpha + \beta x_i + \varepsilon_i$, where the $\varepsilon_i$s are considered random, but they can be estimated by $\hat{\varepsilon}_i = y_i - \hat{\alpha} - \hat{\beta}x_i$, when the data are given.

Relation (3.4.1) can be written in the usual vector form,

$$Y = X\beta + Zu + e, \qquad (3.4.2)$$

where $X$ is a $n \times k$ matrix with elements $x_{ij}$, $\beta = (\beta_1, \beta_2, \ldots, \beta_k)'$, $Z$ is a $n \times r$ matrix with elements $z_{ij}$, $u = (u_1, u_2, \ldots, u_r)'$, and $e = (e_1, e_2, \ldots, e_n)'$.

In most practical cases, the two random vectors $u$ and $e$ are assumed to be independent and normally distributed. Moreover, we let $\text{Var}(e) \equiv R$, and $R = \sigma_e^2 I$, unless specified otherwise. However, the $u_j$s are independent only when all the $u$s are unrelated. This is seldom the case in animal or plant breeding because "good" individuals are used to breed over and over. When an additive genetic model is considered, i.e., $d_i = 0$ in (3.3.13), we have

$$\text{Cov}(u_i, u_j) = \sigma_a^2 \cdot 2\phi(u_i, u_j), \qquad (3.4.3)$$

where $\sigma_a^2$ is the additive variance component and $\phi(u_i, u_j)$ is the kinship coefficient between individuals $i$ and $j$. Let

$$\text{Var}(u) = Eu' = \sigma_a^2 A, \qquad (3.4.4)$$

where the $ij$th entry of matrix $A$ is defined by $a_{ij} = 2\phi(u_i, u_j)$.

Regression equation (3.4.2) can now be written in the linear model form,

$$Y = X\beta + \varepsilon, \quad \text{with} \quad \varepsilon = Zu + e \sim N(0, \sigma_e^2(\gamma ZAZ' + I) \equiv N(0, \sigma_e^2 V), \quad (3.4.5)$$

where $\gamma = \sigma_a^2/\sigma_e^2$. Note that the heritability $h^2$ defined in (3.1.7) is $\gamma/(1 + \gamma)$. In many practical situations, the heritability can be estimated by an animal breeder from experience. In this case, $V$ is completely specified and the likelihood function for the unknown parameters $\beta$ and $\sigma_e^2$ is

$$\ln L = -\frac{1}{2} n\ln(2\pi) - \frac{1}{2}\ln|\sigma_e^2 V| - \frac{1}{2}(Y - X\beta)'(\sigma_e^2 V)^{-1}(Y - X\beta). \quad (3.4.6)$$

Thus, we can estimate $\beta$ and $\sigma_e^2$ using the usual general linear model estimation equation

$$\begin{aligned}
\hat{\beta} &= (X'V^{-1}X)^{-1}X'V^{-1}Y, \\
\hat{\sigma}_e^2 &= (Y - X\hat{\beta})'V^{-1}(Y - X\hat{\beta})/(n - k).
\end{aligned} \quad (3.4.7)$$

In the case where $\gamma$ is unknown, there is no close form solution for the parameters. One can try many $\gamma$'s, consider them as known, and use (3.4.6) to estimate $\beta$ and $\sigma_e^2$ and substitute them back to the likelihood equation (3.4.6). The mle estimate of $\gamma$ is the one that maximizes (3.4.6). The computational aspect will be discussed at the end of this section with a numerical example.

The residuals $\hat{e} = Y - X\hat{\beta}$ cannot be separated into $u$ and $e$ without using the conditional expection in multivariate analysis (see Appendix A.1). Note that $(Y, u)$ are jointly normal in the standard form

$$\begin{bmatrix} Y - X\beta \\ u \end{bmatrix} \sim N\left(0, \begin{bmatrix} \Sigma_{11} & \Sigma_{12} \\ \Sigma_{21} & \Sigma_{22} \end{bmatrix}\right), \quad (3.4.8)$$

where

$$\begin{aligned}
\Sigma_{11} &= E\{(Y - X\beta)(Y - X\beta)'\} = E\{(Zu + e)(Zu + e)'\} = \sigma_a^2 ZAZ' + \sigma_e^2 I \\
\Sigma_{12} &= E\{(Y - X\beta)u'\} = E\{(Zu + e)u'\} = \sigma_a^2 ZA, \\
\Sigma_{21} &= \Sigma_{12}', \\
\Sigma_{22} &= \sigma_a^2 A.
\end{aligned}$$

Thus,

$$\begin{aligned}
E(u|Y) &= \Sigma_{21}\Sigma_{11}^{-1}(Y - X\beta) \\
&= \gamma AZ'V^{-1}(Y - X\beta).
\end{aligned} \quad (3.4.9)$$

In practice, we need to substitute the estimated values $\hat{\beta}$, $\hat{\sigma}_a^2$, and $\hat{\sigma}_e^2$ to get $\hat{E}(u|Y)$. The estimate of $u$ is then

$$\hat{u} = \hat{\gamma}AZ'(\hat{\gamma}ZAZ' + I)^{-1}(Y - X\hat{\beta}). \quad (3.4.10)$$

This $\hat{u}$ can be used to evaluate the breeding value of each individual and is called predictor of $u$, in contrast to $\hat{\beta}$ as an estimator of the parameters $\beta$. It is the best linear unbiased predictor (BLUP) of $u$ in the sense that if one wishes to predict a progeny under the condition $x^*$ and $z^*$, i.e., to predict

$$y^* = x^* \beta + z^* u,$$

then $\hat{y}^* = x^* \hat{\beta} + z^* \hat{u}$ minimizes $E(\hat{w} - y^*)^2$ among all the linear unbiased predictors $\hat{w}$.

How good is the prediction? By the conditional expectation in multivariate distribution, we have

$$E(u|Y) = E[u],$$
$$\text{Var}(u|Y) = \Sigma_{22} - \Sigma_{21}\Sigma_{11}^{-1}\Sigma_{12}', \tag{3.4.11}$$

where the $\Sigma$s are defined in (3.4.8). Note these formulae were derived under the condition that all parameters are known. When some parameters are estimated, (3.4.11) is only an approximation.

The significance of the fixed effect can be evaluated by

$$V(\hat{\beta}) = \sigma_e^2 (X'V^{-1}X)^{-1}$$

and that of the variance components by

$$V\begin{bmatrix} \hat{\sigma}_e^2 \\ \hat{\gamma} \end{bmatrix} \doteq \frac{1}{n}\vartheta^{-1},$$

where $\vartheta$ is the information matrix

$$Inf\begin{bmatrix} \sigma_e^2 \\ \gamma \end{bmatrix} = \frac{1}{2}\begin{bmatrix} n/\sigma_e^4 & \text{tr}V^{-1}ZZ'/\sigma_e^2 \\ \text{tr}V^{-1}ZZ'/\sigma_e^2 & \text{tr}(V^{-1}ZZ'V^{-1}ZZ') \end{bmatrix}. \tag{3.4.12}$$

The derivation can be found in Searle et al. (1992, p.241).

In the literature, the main results of (3.4.7) and (3.4.9) have been presented in the following matrix form called the Hendenson mixed model equation (MME, see Henderson, 1976).

$$\begin{bmatrix} X'R^{-1}X & X'R^{-1}Z \\ Z'R^{-1}X & Z'R^{-1}Z + D^{-1} \end{bmatrix}\begin{bmatrix} \hat{\beta} \\ \hat{u} \end{bmatrix} = \begin{bmatrix} X'R^{-1}Y \\ Z'R^{-1}Y \end{bmatrix}, \tag{3.4.13}$$

where $R = \text{Var}(e)$ and $D = \sigma_a^2 A$. This can be derived by the joint density function of $Y$ and $u$,

$$f(Y,u) = g(Y|u)h(u)$$
$$= C\exp\left[-\frac{1}{2}(Y - X\beta - Zu)'R^{-1}(Y - X\beta - Zu)\right]\exp\left[-\frac{1}{2}u'D^{-1}u\right], \tag{3.4.14}$$

where $C$ is a constant, $h(u)$ and $g(Y|u)$ are, respectively, the density function of $u$ and the conditional probability density of $Y$ given $u$. The MME in (3.4.13) was obtained by maximizing (3.4.14) with respect to $\beta$ and $u$ by differentiation. This method, based on the likelihood equation (3.4.6), is called the ML (maximum likelihood) method.

Another often mentioned estimation method is the **restricted maximum likelihood (REML)** estimation for the variance component. The idea is to eliminate the unknown parameters $\beta$ when setting the likelihood equation. Let $K$ be a $(n-r) \times n$ matrix such that $K'X = 0$, where $r = \text{rank}(X)$. An easy choice of $K$ would be to let

$K' = C'[I - X(X'X)^{-1}X']$ with any $(n - r) \times n$ matrix $C'$. (The solution does not depend on $C$.) Multiplying $K'$ to both sides of (3.4.2), we obtain

$$K'Y = K'X\beta + K'Zu + K'e$$
$$= K'Zu + K'e,$$

or $K'Y \sim N(0, K'VK)$. If we work with $K'Y$, $\beta$ is no longer involved. Both ML and REML are accepted by practitioners, but their answers may not be the same. In fact, they can be quite different when the sample size is small. Interested readers may find more on the REML method and its comparison with ML method in Searle et al. (1992, p.249).

**Example 3.4.1** *The following data were obtained from Elzo (1996). In this table there are 12 offspring from three sires which are half-bothers through the same father, but do not have the same mother. The dams of these offsprings belonged to two herds. The breeding value is the weaning weight (wwt).*

| Animal ID $i$ | Herd ($x_{i1}$) | Sire ($u_j$) | WWT($y_i$) |
|---|---|---|---|
| 4 | 1 | 1 | 248 |
| 5 | 1 | 1 | 256 |
| 6 | 1 | 2 | 296 |
| 7 | 1 | 2 | 282 |
| 8 | 1 | 3 | 265 |
| 9 | 1 | 3 | 274 |
| 10 | 2 | 1 | 260 |
| 11 | 2 | 1 | 252 |
| 12 | 2 | 2 | 300 |
| 13 | 2 | 2 | 285 |
| 14 | 2 | 3 | 295 |
| 15 | 2 | 3 | 290 |

Find the herd effect and the breeding values of the three sires.

**Sol:** If we put the data into (3.4.1), we have only one $x_{i1}$ and three $us$. Since the three sires were half-brothers, we have the $a_{ij}$ in (3.4.4) being, $a_{ij} = 1$ for $i = j$ and $a_{ij} = 0.25$ for $i \neq j$. The $X$ and $Z$ are in simple forms, so there is no need to list them. Thus, with given $\sigma_a^2$ and $\sigma_e^2$, $\hat{\beta}$ can be computed from (3.4.7) and the log likelihood value (3.4.6) can be subsequently computed. A simple search over a large range of $\sigma_a^2$ and $\sigma_e^2$ resulted in $\hat{\sigma}_a^2 = 296.0$ and $\hat{\sigma}_e^2 = 65.0$.

Once the $\sigma^2$s are estimated, the SAS MIXED (see Littell et al., 1996) will produce the $\hat{\beta}$ and $\hat{u}$ with their statistical properties. The estimated breeding values for $u_1$, $u_2$, and $u_3$ are respectively 19.82, 14.46 and 5.36. The two coefficients for herds are 270.16 for Herd 1 and 280.3 for Herd 2. The program and outputs are given in Appendix B.8. $\square$

This example is only for demonstration purpose. The data set is much too small compared to those used in real animal breeding. In practice, the main problem is the large number of animals and the inversion of large matrices. In those cases, it may be unrealistic to estimate $\sigma_e^2$ and $\sigma_a^2$ by trial and error. However, when SAS MIXED was used with different "given" $\sigma_a^2$, the same $\hat{\sigma}_a^2 = 296.0$ maximized its log likelihood SAS outputs. Apparently, if the data set is not too large for SAS MIXED to handle, the trial-and-error method is feasible for the variance component estimation.

When many generations of animals are all put into the model (3.4.1), e.g., when animal $i$ is both parent and offspring, there are two ways to construct the model. We can use either let

$$y_i = \sum_{j=1}^{k} x_{ij}\beta_j + u_m + u_f + e_i, \qquad (3.4.15)$$

where $u_m$ and $u_f$ are his mother and father (referred to later as the parental model), or use

$$y_i = \sum_{j=1}^{k} x_{ij}\beta_j + u_i + e_i, \qquad (3.4.16)$$

where $u_i$ is the animal itself (referred to later as the self-model). In (3.4.16), $u_i$ and $e_i$ are separable because they have different covariance structures. If $u_i$s are all independent like the $e_i$s, then they are not separable. If both layout (3.4.15) and layout (3.4.16) are used in the same data set, then (3.4.15) should be

$$y_i = \sum_{j=1}^{k} x_{ij}\beta_j + \frac{1}{2}u_m + \frac{1}{2}u_f + e_i, \qquad (3.4.17)$$

because the influence is only one half from either parents. Judging from the limited literature I reviewed, both models were well used, but only one of the models was used for one data set. Of course, when the parental model includes all subjects, there is no need to put that 1/2 in the $us$ in (3.4.17), because it can be absorbed into the variance of $\sigma_a^2$. As we mentioned at the beginning of this section, where the parental model was used as an example, it is natural to consider the $us$ as random effects, because they are subject to meiotic variation. To consider a $u_i$ in the self-model as a random effect is less straightforward. The argument seems to have to go back to the randomness of the gametes of its parents or ancestors.

## 3.4.2 Marker Assisted Selection (MAS) Model with a Single Marker

Suppose we suspect that a marker is close to an important gene related to a trait $y$. Then, if A and B are relatives with the same allele, they are likely to have the same gene near the marker. Thus, we may use markers to assist genetic prediction (Van Arendonk et al., 1994). Let

$$y_i = \sum_{j=1}^{k} x_{ij}\beta_j + u_i + v_i^p + v_i^m + e_i, \qquad (3.4.18)$$

where all the notation follows (3.4.1), except

$v_i^p$ = genetic contribution to $y_i$ from the father's side, shown by the marker.

$v_i^m$ = genetic contribution to $y_i$ from the mother's side, shown by the marker.

$u_i$ = other genetic effect not linked to the marker, such as other genes unrelated to this marker.

Apparently, only the additive genetic effects can be considered. In matrix form, (3.4.18) becomes

$$Y = X\beta + Zu + Wv + e, \qquad (3.4.19)$$

where $X$, $Z$, $W$, $e$ are defined in (3.4.2) except

$$Wv = \begin{bmatrix} 1 & 1 & 0 & 0 & \cdots & 0 \\ 0 & 0 & 1 & 1 & \cdots & 0 \\ \cdots & & & & & \\ \cdots & & & \cdot & & \\ & & & & 1 & 1 \end{bmatrix} \begin{bmatrix} v_1^p \\ v_1^m \\ v_2^p \\ v_2^m \\ \vdots \\ v_n^p \\ v_n^m \end{bmatrix} \equiv I_n \otimes [1 \quad 1]v,$$

where $I_n$ is an $n \times n$ identity matrix and $\otimes$ is the Kronecker product in matrices. Moreover, $u$, $v$ and $e$ are assumed to be independent of each other. Let the covariance matrix of $v$ be $G_{(2n \times 2n)}$. We use Fig. 3.4.1 as an example to illustrate how $G$ can be computed up to a proportional constant, (see Lemma 3.3.1 and (3.3.13)). (Remark: This figure is the same as Fig. 3.3.4, except marker information is added.)

1) For individuals with unknown ancestors, $v^p$ and $v^m$ can be randomly assigned. For example, in Fig. 3.4.1, the markers 1, 2 for individual 1, are assigned as $v_1^p = 1$ and $v_1^m = 2$. It is easy to see that the submatrix $[G_{4 \times 4}]$ is

$$[G_{4 \times 4}]_2 = \begin{array}{c} \\ v_1^p \\ v_1^m \\ v_2^p \\ v_2^m \end{array} \begin{array}{c} \begin{array}{cccc} v_1^p & v_1^m & v_2^p & v_2^m \end{array} \\ \begin{pmatrix} 1 & 0 & 0 & 0 \\ 0 & 1 & 0 & 0 \\ 0 & 0 & 1 & 0 \\ 0 & 0 & 0 & 1 \end{pmatrix} \end{array}. \qquad (3.4.20)$$

To add the next individual, 3, the submatrix $6 \times 6$ is

$$[G_{6 \times 6}]_3 = \begin{bmatrix} [G]_2 & [G]_2 & \begin{array}{c} \begin{array}{cc} v_3^p & v_3^m \end{array} \\ \begin{pmatrix} 1-r & 0 \\ r & 0 \\ 0 & 1-r \\ 0 & r \end{pmatrix} \end{array} \\ \text{Symmetry} & & \begin{bmatrix} 1 & 0 \\ 0 & 1 \end{bmatrix} \end{bmatrix}, \qquad (3.4.21)$$

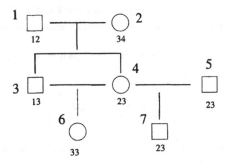

Figure 3.4.1: Pedigree for MAS G-matrix computation. The founding ancestor geno-type assignments are: $v_1^p = 1$, $v_1^m = 2$, $v_2^p = 3$, $v_2^m = 4$, $v_5^p = 2$, and $v_5^m = 3$.

where $[G]_2$ is defined in (3.4.20), and $r$ is the recombination fraction between the marker and the gene. (3.4.21) is apparent because $v_1^p$ and $v_3^p$ are IBD with probability $1 - r$. In general, from $j$th individual to $(j+1)$th individual,

$$[G_{2(j+1)\times2(j+1)}]_{j+1} = \begin{bmatrix} [G_{2j\times2j}]_j & [G]_j S_j \\ \text{symmetry} & \begin{pmatrix} 1 & 0 \\ 0 & 1 \end{pmatrix} \end{bmatrix}, \tag{3.4.22}$$

where $S_j$ is a $2j \times 2$ matrix with all 0 entries except a $\binom{r}{1-r}$ or $\binom{1-r}{r}$ in each column indicating the relationship between $j$ and its parents. Whether to put $\binom{r}{1-r}$ or $\binom{1-r}{r}$ in the $S_j$ matrix depends on the marker relation. For example, for individual 4,

$$S_4 = \begin{matrix} & v_4^p \quad\; v_4^m \\ \begin{pmatrix} r & 0 \\ 1-r & 0 \\ 0 & 1-r \\ 0 & r \\ 0 & 0 \\ 0 & 0 \end{pmatrix} \end{matrix}$$

and for individual 6,

$$S_6 = \begin{matrix} & v_6^p \quad\; v_6^m \\ \begin{pmatrix} 0 & 0 \\ 0 & 0 \\ 0 & 0 \\ 0 & 0 \\ r & 0 \\ 1-r & 0 \\ 0 & r \\ 0 & 1-r \\ 0 & 0 \\ 0 & 0 \end{pmatrix} \end{matrix}.$$

$S_7$ is different because there is no way to trace the 2 and 3. Thus it is $\binom{0.5}{0.5}$ for $v_7^p$ in columns 9, 10 and $\binom{0.5}{0.5}$ for $v_7^m$ in columns 7, 8. It can also be proven that $G$ is positive definite.

Parameter estimation for (3.4.19) is similar to (3.4.2), except the variance matrix now is

$$\operatorname{Var}\begin{bmatrix} u \\ v \\ e \end{bmatrix} = \sigma_e^2 \begin{bmatrix} \gamma A & 0 & 0 \\ 0 & \delta G & 0 \\ 0 & 0 & I \end{bmatrix},$$

where $\gamma$ is defined as before and $\delta$ is the variance ratio between the gene and $e$. The solution to the linear equation is

$$\begin{bmatrix} X'X & X'Z & X'W \\ Z'X & Z'Z + (\gamma A)^{-1} & Z'W \\ W'X & W'Z & W'W + (\delta G)^{-1} \end{bmatrix} \begin{bmatrix} \hat{\beta} \\ \hat{u} \\ \hat{v} \end{bmatrix} = \begin{bmatrix} X'Y \\ Z'Y \\ W'Y \end{bmatrix}, \qquad (3.4.23)$$

when $\gamma$ and $\delta$ are known. Otherwise, we have to maximize a likelihood equation similar to (3.4.6). The detailed derivation is left as an exercise.

### 3.4.3  MAS With Multiple Markers

It is easy to extend (3.4.18) to multiple markers. Let there be $t$ markers. Define

$v_{ik}^p$ = genetic contribution to $y_i$ from the father's side related to marker $k$, and define $v_{ik}^m$ similarly for the mother's side.

Then (3.4.18) becomes

$$y_i = \sum_{j=1}^{k} x_{ij}\beta_j + u_i + \sum_{k=1}^{t}(v_{ik}^p + v_{ik}^m) + e_i, \qquad (3.4.24)$$

or in matrix form

$$Y = X\beta + Zu + \sum_{k=1}^{t} W_k v_k + e, \qquad (3.4.25)$$

where $v_k$ is defined the same as the $v$s in (3.4.19). The solution for (3.4.25) is a simple extension of (3.4.23) using a separate covariance matrix $G_k$ for every $v_k$, $k = 1, 2, \ldots, t$. There are, of course, more variance ratios to be estimated.

When the markers become even more numerous, it is better not to use only the relationship of the gene to individual markers but to consider the gene transition using two neighboring markers, as shown in Fig. 3.4.2. Without loss of generality, let $Q_1$ be an influential gene to the trait. We need to trace whether $Q_1$ is passed through meiosis, given the knowledge of the markers. Note that the distance between $M_1$ and $M_2$ is known when the markers are identified. Assume the markers are highly polymorphic,

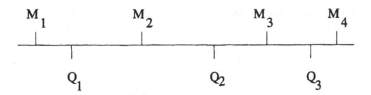

Figure 3.4.2: The relation between known position markers $M_i$ and trait loci $Q_i$ on a chromosome.

i.e., we assume the line of descent for any marker can be traced without error. Let the two gametes be shown in Fig. 3.4.3(a), i.e. the gametes are sandwiched as $(1v_p1)/(2v_m2)$. Note that the two 1s are not the same; they are alleles at two different loci. The four possible haplotypes from the father are shown in Fig. 3.4.3(b), where the $v_{ij}$ can be $v_p$ or $v_m$ depending on how the recombination shuffled it during meiosis.

Let the crossover probability between $M_1$ and $Q_1$ be $\theta_1$ and that between $Q_1$ and $M_2$ be $\theta_2$. Then the crossover between $M_1$ and $M_2$ is $\theta_2(1 - \theta_1) + \theta_1(1 - \theta_2) \simeq \theta_1 + \theta_2 \equiv r$, assuming no double-crossing. Let $\theta_1/r = p_1, \theta_2/r = 1 - p_1 = q_1$. Thus, $Pr\{v_{11} = v_p\} = Pr\{v_{22} = v_m\} = 1$ by the no double crossover assumption. Also,

$$Pr\{v_{12} = v_p\} = Pr\{v_{21} = v_m\} = q_1 \quad \text{and} \quad Pr\{v_{21} = v_m\} = Pr\{v_{21} = v_p\} = p_1.$$

Now the genetic relationship between generations as demonstrated by the two markers can still be written as (3.4.22) with

$$S_j = [s_p \ s_m],$$

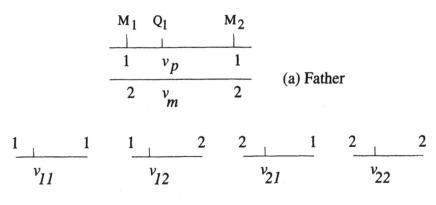

(a) Father

(b) Son from father's side

Figure 3.4.3: Four possible relations between a parent and offspring gametes during meiosis; (a) father's gamete; (b) son's haplotype from the father's side.

where $i$ is the loci index and $j$ is the individual index as the $j$ in (3.4.22). Similarly, $s_p$ (or $s_m$) have 0 entries except the two columns related to his father, i.e.,

$$\begin{bmatrix} s_{11}^p \\ s_{22}^p \end{bmatrix} = \begin{cases} \begin{pmatrix} 1 \\ 0 \end{pmatrix} & \text{if we see markers 11} \\ \begin{pmatrix} q_1 \\ p_1 \end{pmatrix} & \text{if we see markers 12} \\ \begin{pmatrix} p_1 \\ q_1 \end{pmatrix} & \text{if we see markers 21} \\ \begin{pmatrix} 0 \\ 1 \end{pmatrix} & \text{if we see markers 22.} \end{cases}$$

According to Goddard (1989), $p_i$, as well as $\sigma_k^2$ for $G_k$, have to be estimated. But with so many parameters, the computation can be difficult. A simpler case is in a back-crossing breeding with the assumption that the $p_i$'s are uniformly distributed between the two markers. Without loss of generality in an additive model, we may let $v_p$ and $v_m$ be $-v$ and $v$, respectively, in Fig. 3.4.4. For markers with configurations $[1, -1]$ or $[-1, 1]$,

$$\begin{aligned} pv_p + qv_m &= -pv + qv = (1 - 2p)v, \\ qv_p + pv_m &= (2p - 1)v. \end{aligned} \tag{3.4.26}$$

In this case,

$$s_{11}^p = Z \begin{bmatrix} (1 - p)v \\ pv \end{bmatrix},$$

where $Z = [-1\ -1]$ for alleles $(-1, -1)$; $[-1\ 1]$ for alleles $(-1, +1)$; $[1\ -1]$ for alleles $(+1, -1)$ and $[1\ 1]$ for alleles $(+1, +1)$. It can be shown that, with the uniform distribution assumption of $p$,

$$\text{Var} \begin{bmatrix} (1 - p)v \\ pv \end{bmatrix} = \begin{bmatrix} 1/3 & 1/6 \\ 1/6 & 1/3 \end{bmatrix} \sigma_v^2. \tag{3.4.27}$$

The proof is left as an exercise.

For consecutive markers, we still use Fig. 3.4.4 as an example. In this figure, the father's genotype is given and so are the offspring's markers. The offspring's genotype $(g_1, g_2, g_3)$ are assumed to be random variables. Goddard (1992) devised a very efficient way to represent the effect of $v_k$ in (3.4.25) by

$$g_1^{2p} + g_2^{2p} + g_3^{2p} = [m_1\ m_2\ m_3\ m_4] \begin{bmatrix} q_1 g_1^{1p} \\ p_1 g_1^{1p} + q_2 g_2^{1p} \\ p_2 g_2^{1p} + q_3 g_3^{1p} \\ p_3 g_3^{1p} \end{bmatrix}, \tag{3.4.28}$$

or, in general,

$$g_i^{2p} = q_i m_i g_i^{1p} + p_i m_{i+1} g_i^{1p},$$

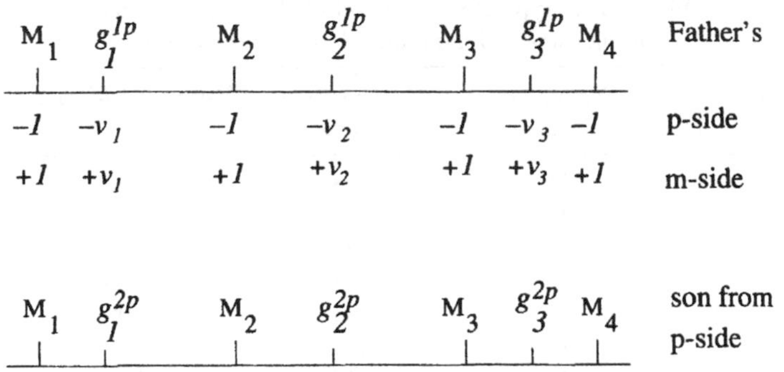

Figure 3.4.4: Notation for Goddard's MAS model. The father's two haploids are one from his mother's side and one from his father's side.

and, if the marker $m_i$ and $m_{i+1}$ is replaced by $\pm 1$, we have the correct $g_i^{2p}$. For example, $[m_i, m_{i+1}] = [-1, 1]$ implies

$$g_2^{2p} = -q_i g_i^{1p} + p_i g_i^{1p} = (2p_i - 1) g_i^{1p},$$

which agrees with (3.4.26). Thus (3.4.26) can be written in the form of (3.4.2)

$$Y = X\beta + \Sigma Z_i g_i + e. \tag{3.4.29}$$

The covariance matrix can be obtained by (3.4.27) with the independence assumption among the loci. (3.4.10) is now directly applicable to finding the prediction value

$$E(u_i + \sum_{k=1}^{t} (v_{ik}^p + v_{ik}^m)|Y)$$

after all the parameters have been estimated.

## 3.4.4 A Simulation Study

A simulation study is probably one of the best ways to check whether a statistical method works under the given assumptions. We use the study of Zhang and Smith (1992) as an example.

A. Scope of the simulation:

  A.1 There are 20 chromosomes, each 1 Morgan in length.
  A.2 On each chromosome, 5 QTL and 5 markers are randomly assigned. This makes a total 100 QTL and 100 markers.
  A.3 Using the notation in (3.1.2), the genetic effects of QTL are assigned, without loss of generality, by letting $d = 0$ and $a = \mu_j$ for the $j$th gene. Also, HWE is assumed with $p = q = 0.5$ for all the genes. Thus, $\sigma_g^2 = 0.5\mu_j^2$ for the $j$th gene

Table 3.4.1: A typical marker and QTL assignment on a
chromosome in Zhang and Smith's simulation.

| $\mu_i$ $(\sigma^2 = 1)$ | Marker or QTL | Map position (in Morgan) | Estimated MQTL |
|---|---|---|---|
|  | M | 0.08 | 0.00 |
|  | M | 0.19 | 0.04 |
| 0.15 | Q | 0.22 |  |
|  | M | 0.23 | 0.10 |
| −0.05 | Q | 0.25 | 0.02 |
|  | Q | 0.39 |  |
|  | M | 0.49 | 0.04 |
| 0.02 | Q | 0.73 |  |
| −0.06 | Q | 0.95 |  |
|  | M | 0.99 | −0.06 |

From Zhang and Smith (1992), copyright ©by Springer-Verlag, Inc. and by
the authors. Reprinted with permission.

(see 3.1.5). The magnitudes of $\mu_i$s are collected from a zero mean normal
distribution and normalized using

$$0.5 \sum_{j=1}^{100} \mu_j^2 = h^2 \sigma^2,$$

where $\sigma^2$ is the variance for the trait and $h^2$ is the heritability. More precisely,
the phenotype model is

$$Y_i = \mu + \sum_{j=1}^{100} (v_{ij}^p + v_{ij}^m) + e_i,$$

where $v_{ij}^p$ and $v_{ij}^m$ are the effects of the two alleles at locus $j$ for individual $i$, and
$e_i \sim N(0, (1 - h^2)\sigma^2))$. In the simulation, $\sigma^2 \equiv 1$ and $h^2 = 0.5, 0.25, 0.1$.

A typical assignment on one chromosome is given in Table 3.4.1. The last
column will be discussed later. Note that the position of Q is assumed to be
unknown in data analysis, but its importance should be revealed by the
adjacent markers.

A.4 At $F_0$ generation, all QTL and markers are homozygous, each with two alleles
($AA$ or $aa$, $MM$ or $mm$). There are two individuals at $F_0$, one is the complement of the other, i.e., if one is $AA$ then at the same locus the other is $aa$. $F_1$
are the offspring between the two $F_0$ individuals. Hence all the individuals in
$F_1$ have the same genotype $Aa$ and $Mm$. The population is then allowed to
randomly mate from 2 to 20 generations with crossover between genes and
markers according to their distances.

B. Analysis and breeding selection

B.1 Three models are used for the data analysis and to estimate the breeding value.
1. BLUP without marker, i.e., by (3.4.2).
2. Goddard's MAS model with

$$Y = \mu + Zv + e,$$

where $Zv$ represent the marker-assisted model (see (3.4.29)).
3. A selection method (COMB) that combines BLUP and MAS weighted by their relative variance was proposed by Lande and Thompson (1990). The idea was to eliminate inefficient markers and rerun the analysis with fewer number of markers. This method should work very well because markers not adjacent to any important genes simply add noise in the data analysis.

B.2 One generation, such as $F_2$, $F_5$, $F_{10}$ or $F_{20}$, was picked as the starting point of the breeding experiment. At the beginning of the breeding process, 500 males and 500 females were randomly selected for breeding value prediction. The best thirty males and thirty females were chosen to produce 500 males and females of the next generation, according to the three models; MAS, BLUP, and COMB. This selection continued for 10 generations. The last column on Table 3.4.1 estimates the $vs$ in (3.4.24) at $F_2$.

B.3 The simulation was repeated 30 times and the results were given in Table 3.4.2. The values were the sample mean ± one sample standard deviation based on 30 replicated simulations. The $F_2$ and $F_{10}$ in the parentheses were the

Table 3.4.2: Performance comparisons of the three selection methods.

| $h^2$ | Generation | BLUP | MAS($F_2$) | COMB($F_2$) | MAS($F_{10}$) | COMB($F_{10}$) |
|---|---|---|---|---|---|---|
| 0.10 | 1 | 0.00±0.06 | 0.00±0.05 | 0.00±0.05 | 0.00±0.06 | 0.00±0.06 |
| | 2 | 1.04±0.24 | 0.93±0.10 | 1.17±0.17 | 0.35±0.13 | 1.04±0.15 |
| | 3 | 1.62±0.25 | 1.45±0.10 | 1.93±0.17 | 0.45±0.15 | 1.64±0.17 |
| | 5 | 2.27±0.26 | 2.08±0.11 | 2.87±0.18 | 0.73±0.18 | 2.49±0.18 |
| | 10 | 3.59±0.27 | 2.61±0.16 | 3.95±0.16 | 1.03±0.20 | 3.68±0.20 |
| 0.25 | 1 | 0.00±0.10 | 0.00±0.08 | 0.00±0.08 | 0.00±0.09 | 0.00±0.10 |
| | 2 | 1.10±0.30 | 1.03±0.14 | 1.42±0.21 | 0.56±0.18 | 1.20±0.23 |
| | 3 | 1.90±0.31 | 1.62±0.17 | 2.18±0.23 | 0.83±0.22 | 2.00±0.24 |
| | 5 | 2.81±0.31 | 2.38±0.24 | 3.60±0.29 | 1.11±0.26 | 3.01±0.25 |
| | 10 | 4.11±0.34 | 3.03±0.32 | 4.42±0.26 | 1.52±0.39 | 4.21±0.27 |
| 0.50 | 1 | 0.00±0.17 | 0.00±0.12 | 0.00±0.12 | 0.00±0.13 | 0.00±0.13 |
| | 2 | 1.30±0.41 | 1.24±0.22 | 1.50±0.25 | 0.70±0.27 | 1.27±0.20 |
| | 3 | 2.13±0.40 | 1.93±0.25 | 2.41±0.29 | 0.98±0.33 | 2.24±0.22 |
| | 5 | 3.12±0.43 | 2.84±0.32 | 3.85±0.26 | 1.37±0.46 | 3.29±0.29 |
| | 10 | 4.81±0.42 | 3.86±0.41 | 5.09±0.29 | 1.85±0.60 | 4.82±0.32 |

From Zhang and Smith (1992), copyright ©by Springer-Verlag, Inc. and by the authors. Reprinted with permission.

starting generations. Note that BLUP has only one column because it is independent of the starting generation.

From this table we see that the breeding value increases when a selection method was applied. The pure BLUP model did very well, the MAS model did poorly, but the combined method seemed to give the best prediction.

There is a close relation between the MAS model in this section and the polygene model in §3.2.5. We expect a more unified approach to both models in the future.

# Exercise 3

3.1 What changes need to be made if the parents markers are known in (3.1.9)? Use the following three examples to discuss the changes. (i) Parents' marker types are (1,2) (3,4). (ii) Parents' marker types are (1,2) (1,3). (iii) Parents' marker types are (1,1) (1,3).

3.2 Assume that in Table 3.1.1, all the $f$s $\mu$s and $\sigma$s are known and all the three $\sigma$s are the same. (i) Suppose $n = 2$ individuals are randomly sampled in the population. Find the conditional probability of the individual with the higher trait value being $BB$, $Bb$, and $bb$. (ii) Describe a numerical method that can compute this probability for any $n$. (iii) Also, use Table 3.1.1 to derive the relation between $f(x, y|\pi_t)$ and $f(x, y|g_t)$ in evaluating (3.1.20).

3.3 Derive (3.2.8) and (3.2.9).

3.4 Derive (3.2.19).

3.5 Explain why heritability cannot be independent of the population structure. In other words, heritability is a local phenomenon.

3.6 Derive $\Pr\{g_2 = B|m = (11/12)\}$ in Table 3.2.1.

3.7 Derive the three values $\rho$, E[W], and E[B] for the case MZA in Table 3.3.1.

3.8 Scarr and Carter-Saltzman (1983, Table 7.3, p. 260) summarized several twin studies on genetics and IQ scores in the following table.

|                                    | Correlation | Number of Pairs |
|------------------------------------|-------------|-----------------|
| MZT                                | 0.857       | 526             |
| DZT                                | 0.534       | 517             |
| MZA                                | 0.741       | 69              |
| Sibs reared together               | 0.545       | 1671            |
| Unrelated children reared together | 0.376       | 259             |

Estimate the heritability and find a 95% confidence interval for your estimate.

3.9 (i) Show that $Cov(X_1, X_2) = \sigma_a^2/2$, where 1 and 2 are father and son when all the marriages were formed by unrelated persons. (ii) One intuitive indicator of herit ability is to study the relation between parents and children. Past large sample studies have shown that the correlation between the heights of father and son (see e.g., Freedman et al., 1980, p. 160) is approximately 0.5. What is a reasonable estimate of heritability (in what sense)? Is your answer reasonable? Why?

3.10 One important observation in heredity is the regression toward the mean, i.e., a tall father is likely to have a son shorter than himself and vice versa. Try to formulate this phenomenon and prove it mathematically.

3.11 Use the computational algorithm decribed in Example 3.3.4 to show that $D = 2\Phi$ is a positive semi-definite. Hence, it can be used as a variance-covariance matrix.

3.12 Derive (3.4.23). (Hint: Consider (3.4.18) as a special case of (3.4.2)).

3.13 Derive (3.4.27).

3.14 The following 10 sibpairs (Fig. E3.14) are obtained to study the possible linkage between the three markers and a quantitative gene. The marker alleles are denoted consecutively in the pedigrees and the trait values of the sibs are denoted by darker numbers. The recombination fractions between both consecutive markers are 20 cM. Use *MAPMAKER/SIBS* (see Appendix B.6) to do the linkage analysis.

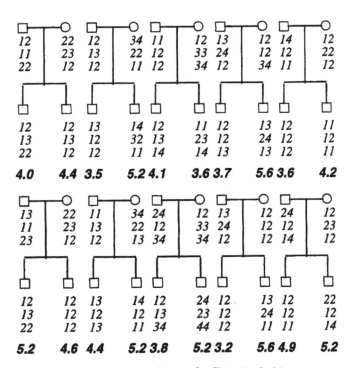

Figure E3.14: Pedigrees for Exercise 3. 14.

# Chapter 4

# Special Topics

## 4.1 Small Area Gene Mapping by Linkage Disequilibrium

In human pedigree analysis, the gene cannot be pinpointed to a very small area, say less than 1 cM. The reason is that when the recombination fraction is very small, crossover can seldom be observed in a few generations. One way that we may be able to increase the sample size is to jump from families (or siblings) to subpopulation. For example, suppose a group of people has been settled in one geographic area for 600 years, and suppose the disease originated from a single mutation 600 years ago. Moreover, at the time of mutation, a certain marker was very close to the disease gene, say 0.1 cM (100 kbp) away. Then, after 600 years, or approximately 20 generations, the same marker should still be linked to the mutant gene because of the small (0.001) crossover probability. Any pedigree is unlikely to trace back 20 generations, but the linkage between the disease and this particular marker can still be seen from the population in this area. If one allele from a marker locus appears more often in affected individuals than in unaffected individuals, then we may suspect that this marker is related, and perhaps very close, to the disease gene. Table 4.1.1 gives the frequency tables for four markers and cystic fibrosis (CF, see a description of this disease in Example 2.2.3) patients and their controls (from Estivill et al., 1987).

These are the well-known $2 \times 2$ contingency tables. Either Fisher's exact test or the asymptotic chi-square test (see Appendix A.4) can be used to test the independence of the disease and the markers. More precisely, let the table be laid out in general as is Table 4.1.2.

From this table, the chi-square test for independence between the row and column is

$$Q = \sum_{i=a}^{A} \sum_{j=b}^{B} \frac{(n_{ij} - n_i \times n_j/n)^2}{n_i \times n_j/n} = \frac{n(n_{AB}n_{ab} - n_{Ab}n_{aB})^2}{n_A n_a n_B n_b} \sim \chi_1^2. \tag{4.1.1}$$

Table 4.1.1: Contingency tables for CF patients and their controls with 4 markers. The markers are shown at the bottom. The allele numbers 1 and 2 are at the very left of each table and chi-square values are Q.

| | CF | Normal | | CF | Normal | | CF | Normal | | CF | Normal |
|---|---|---|---|---|---|---|---|---|---|---|---|
| 1 | 73 | 56 | 1 | 90 | 89 | 1 | 195 | 93 | 1 | 1 | 38 |
| 2 | 61 | 75 | 2 | 24 | 23 | 2 | 18 | 109 | 2 | 71 | 28 |

| pJ3.11 Msp1 | pmetD TagI | PXV-2C TagI | pCS.7 HhaI |
|---|---|---|---|
| Q = 3.2 | Q = 0.01 | Q = 103.0 | Q = 54.0 |

Table 4.1.2: Notation for a $2 \times 2$ contingency table

| | $A$ | $a$ | Total |
|---|---|---|---|
| $B$ | $n_{AB}$ | $n_{aB}$ | $n_B$ |
| $b$ | $n_{Ab}$ | $n_{ab}$ | $n_b$ |
| Total | $n_A$ | $n_a$ | $n$ |

From the chi-square table values, $\chi^2_{1,0.05} = 4.84$ and $\Pr\{\chi^2_1 > 25\} = 2.67 \times 10^{-7}$, Table 4.1.1 indicates that the CF gene is highly linked to the PXV-2C and PCS.7 markers. The power of this chi-square test can be easily computed when the alternative is specified (see Appendix A.4).

Although this method is simple, it has two stringent requirements. (i) The marker locus and the disease locus have to be extremely close. Otherwise, the disease gene and markers should have reached linkage equilibrium i.e., the marker and the gene are independent for two unrelated persons. And (ii), the mutation has to be rare. Otherwise, the disease gene may associate with many markers in many occasions of mutation, and this method will not work. Lander and Schork (1994) have stated some other concerns about using population association to locate a gene.

In practice, locating a gene by linkage disequilibrium (or by association, see Weil 1996 p. 112) is effective only when the gene's location is approximately known. Obviously, due to the extreme closeness requirement between the marker and the gene, it is impossible to do a genome-wide search. For example, with 1 cM dense markers, we need 3000 marker loci for genome-wide search. It is an unbearable cost if all of these markers need to be probed in a large number of individuals. Moreover, the false-alarm rate can be very high. Thus, we need to have a good idea of the gene's location. This is exactly what the linkage analysis by pedigree or by affected-sib-pairs can tell us. Jorde et al. (1994) did some experiment on human chromosome 5 and found that the best distance for linkage disequilibrium to detect is 50−500 kb, or $\theta = 0.0005$−$0.005$, a perfect complement for the linkage analysis. The current gene hunting are usually start with linkage analysis to pinpoint a gene within 5 cM−1 cM interval and then use

Table 4.1.3: Notation for the joint distribution of two loci $u$ (with alleles $A$ and $a$) and $v$ (with alleles $B$ and $b$). $p_{ij}$ = gametic frequency for genotype $ij$ and $p_i$ = allele frequency for allele $i$.

|  | $A$ | $a$ | Total |
|---|---|---|---|
| $B$ | $p_{AB}$ | $p_{aB}$ | $p_B$ |
| $b$ | $p_{Ab}$ | $p_{ab}$ | $p_b$ |
| Total | $p_A$ | $p_a$ | 1 |

linkage disequilibrium to do the fine tuning. Examples will be given in the latter part of this section.

To put the linkage disequilibrium method more precisely, we will first change (4.1.1) into a more easily interpretable form from a genetic viewpoint. Let the $A$ and $a$ be the alleles at the disease locus and $B$ and $b$ be the alleles at the marker locus. Thus, the $A/a$ phenotypes in Table 4.1.2 have to be changed to alleles $A/a$ for another marker. Let $n_{AB}, n_{Ab}, n_{aB}, n_{ab}$ be the gametic counts of $AB, Ab, aB, ab$ respectively. (See §1.1 for the definition of gamete.) Let the theoretical frequencies of Table 4.1.2 be listed in Table 4.1.3.

Then, under linkage equilibrium $p_{AB} = p_A p_B$, but in general, we define

$$D = p_{AB} - p_A p_B \qquad (4.1.2)$$

as the **disequilibrium coefficient**. After (4.1.2), the other discrepancies can be derived as are those in Table 4.1.4.

We can use (4.1.1) to test $H_0 : D = 0$ indirectly, but we can test $D$ directly from $n$ gametes by letting

$$\hat{p}_{ij} = n_{ij}/n, \quad \hat{p}_i = n_i/n, \quad \hat{p}_j = n_j/n, \quad \text{and} \quad \hat{D} = \hat{p}_{AB} - \hat{p}_A \hat{p}_B, \qquad (4.1.3)$$

where $i = A$ or $a$, $j = B$ or $b$. It can be shown (Weir, 1996, p.113) that asymptotically,

$$E[\hat{D}] = \frac{n-1}{n} D, \quad \text{and}$$

$$\text{Var}[\hat{D}] = \frac{1}{n}[p_A(1-p_A)p_B(1-p_B) - (1-2p_A)(1-2p_B)D - D^2],$$

Table 4.1.4: The discrepancies of $p_{ij} - p_i p_j$ in linkage disequilibrium.

| Gamate | Under equilibrium | Actual freq. | Discrepancy |
|---|---|---|---|
| $AB$ | $p_A p_B$ | $p_{AB}$ | $D = p_{AB} - p_A p_B$ |
| $Ab$ | $p_A p_b$ | $p_{Ab}$ | $-D = p_{Ab} - p_A p_b$ |
| $aB$ | $p_a p_B$ | $p_{aB}$ | $-D = p_{aB} - p_a p_B$ |
| $ab$ | $p_a p_b$ | $p_{ab}$ | $D = p_{ab} - p_a p_b$ |

and

$$\frac{\sqrt{n}\hat{D}}{\{\hat{p}_A(1-\hat{p}_A)\hat{p}_B(1-\hat{p}_B)\}^{1/2}} \sim N(0,1), \quad \text{or}$$

$$z^2 \equiv \frac{n\hat{D}^2}{\hat{p}_A(1-\hat{p}_A)\hat{p}_B(1-\hat{p}_B)} \sim \chi_1^2. \tag{4.1.4}$$

Note that $n$ is the total number of gametes, or twice the number of individuals if no gametes are missing in the experiment. Note that (4.1.4) is true even there are more than two alleles. When there are two alleles, the theoretical value of (4.1.4) without the $n$ is called the **standard equilibrium coefficient**

$$r = \frac{D}{\sqrt{p_A p_a p_B p_b}}. \tag{4.1.5}$$

It turns out, under certain conditions, that

$$r^2 \approx \frac{1}{1+4N\theta}, \tag{4.1.6}$$

where $\theta$ is the recombination fraction between loci $u$ and $v$ and $N$ is the population size in a closed community. The meaning of $N$ will become clearer in the following derivation. This formula has been used to roughly estimate $\theta$ (e.g., Estivill et al., 1987 and Jorde et al., 1994). Here we follow the derivation by Sved (1971), although similar results had been discovered by Hill and Robertson (1968) and Kimura and Crow (1964).

Consider the following sampling scheme: First randomly sample a gamete from a population and then take another gamete from the population so that the second gamete has the same allele as the first at position $u$. Let $P_H$ be the probability that the alleles at position $v$ is also homozygous. Let the genotypes of the gametes be denoted by $g_1 = AB$, $g_2 = Ab$, $g_3 = aB$, and $g_4 = ab$, and their population frequencies be $p_{AB}$, $p_{Ab}$, $p_A$ etc. as defined in Table 4.1.3. Then

$$P_H = \sum_{i=1}^{2} p(g_i) \cdot p(g_i|u=A) + \sum_{i=3}^{4} p(g_i) \cdot p(g_i|u=a),$$

where the order of the products in the summation represents the order of the sampling procedure. Thus,

$$\begin{aligned} P_H &= p_{AB} \cdot \frac{p_{AB}}{p_A} + P_{Ab} \cdot \frac{p_{Ab}}{p_A} + p_{aB}\frac{p_{aB}}{p_a} + p_{ab}\frac{p_{ab}}{p_a} \\ &= \frac{p_{AB}^2 + p_{Ab}^2}{p_A} + \frac{p_{aB}^2 + p_{ab}^2}{p_a} \\ &= \frac{(p_A p_B + D)^2 + (p_A p_b - D)^2}{p_A} + \frac{(p_a p_B - D)^2 + (p_a p_b + D)^2}{p_a} \\ &\doteq p_B^2 + p_b^2 + \frac{2D^2}{p_A p_a}. \end{aligned} \tag{4.1.7}$$

On the other hand, let $I = Pr\{$Any two randomly selected gametes are IBD at $u$ and no recombination has ever happened between loci $u$ and $v\}$. Then,

$$P_H = I + (1 - I)(p_B^2 + p_b^2). \tag{4.1.8}$$

Note that $(1 - I)$ contains the probability of two events; one is that the two haploids are unrelated and, in this case, the probability of being homozygous at $v$ is $p_B^2 + p_b^2$. The other event is that the two haploids are from a common ancestor but underwent crossover between loci $u$ and $v$. Thus, the chance of the allele received at $v$ locus after crossover with some other gamete is $p_B$ for $B$ and $p_b$ for $b$, i.e., the allele at $v$ locus has the same distribution as the general population. Together with the original allele $v$ which has probability $p_B$ for $B$ and $p_b$ for $b$, the probability of being homozygous at $v$ is also $p_B^2 + p_b^2$. Thus, we have (4.1.8).

Equating (4.1.7) and (4.1.8), we have

$$I = \frac{D^2}{p_A p_a p_B p_b} = r^2. \tag{4.1.9}$$

Suppose there are $N$ individuals in a closed community and the the matings are random. Then it can be seen that the relation between the the $n$th generation $I_n$ of (4.1.8) and the next generation $I_{n+1}$ is

$$I_{n+1} = \frac{1}{2N}(1 - \theta)^2 + (1 - \frac{1}{2N})(1 - \theta)^2 I_n, \tag{4.1.10}$$

where $(1 - \theta)^2$ means there is no recombination between loci $u$ and $v$ in two chromosomes randomly picked from a possible $2N$ chromosomes after meiosis, $1/(2N)$ means the chance the same chromosome is sampled twice (using a *with replacement* approximation), and $(1 - 1/(2N))I_n$ means the chance of getting two unrecombined IBD chromosomes from the previous generation. Eq. (4.1.10) is a simple first order difference equation. But we are only interested in its limiting distribution. If we let $I = \lim_{n \to \infty} I_n$ and substitute it into both sides of (4.1.10), we have

$$I = \frac{(1 - \theta)^2/(2N)}{1 - (1 - 1/(2N))(1 - \theta)^2} \approx \frac{1}{1 + 4N\theta}.$$

By (4.1.9), (4.1.6) is proven. $\qquad\qquad\qquad\qquad\qquad\qquad\qquad\qquad\square$

Note that the assumptions in proving (4.1.6), such as random mating in a constant community size of $N$ over many generations, can never be a good approximation. Thus (4.1.6) can only be used as a very rough estimate of $\theta$. For a more rigorous approach, see Hill and Weir (1994).

**Example 4.1.1** *Table 4.1.5 (from Lucassen et al., 1993) shows the possible linkage between a diabetes gene and markers around INS in chromosome 11p (Julier et al., 1991) and Fig. 4.1.1 (Lucassen et al., 1993).*

In this table, the symbol ++ means a homozygous locus with a marker that may be associated with the recessive diabetes gene. Thus, the symbols +/- and -/- denote

Table 4.1.5: Association tests between a diabetes gene and the VNTRs near INS region.

| Polymorphism | Controls | | Diabetics | | RR | 95% | $\chi^2$ |
|---|---|---|---|---|---|---|---|
| | ++ | +/- or -/- | ++ | +/- or -/- | | *C.I.* | |
| TH microsatellite (122 bp) | 13 (14%) | 79 (86%) | 8 (5%) | 143 (95%) | 0.35 | 0.1–0.8 | 5.6 |
| −4127 Pstl (a) | 33 (35%) | 62 (65%) | 41 (27%) | 110 (72%) | 0.7 | 0.4–1.2 | 1.6 |
| −2733 A/C (A) | 36 (38%) | 60 (62%) | 109 (70%) | 47 (30%) | 3.8 | 2.2–6.5 | 25.5 |
| −2221 Mspl (p) | 53 (55%) | 43 (45%) | 131 (84%) | 25 (16%) | 4.2 | 2.3–7.5 | 25.0 |
| −365 VNTR[a] (cl l) | 42 (45%) | 54 (55%) | 120 (78.5) | 34 (22%) | 4.5 | 2.6–7.8 | 30.3 |
| −23 Hphl[a] (p) | 42 (45%) | 54 (55%) | 122 (78%) | 34 (22%) | 4.5 | 2.6–7.9 | 31.0 |
| +805 DraIII (a) | 52 (54%) | 44 (46%) | 129 (83%) | 27 (17%) | 4.0 | 2.3–7.1 | 23.9 |
| +1127 Pstl[b] (p) | 53 (55%) | 43 (45%) | 131 (84%) | 25 (16%) | 4.2 | 2.3–7.5 | 25.0 |
| +1140 A.C[a] (C) | 42 (45%) | 54 (55%) | 122 (78%) | 34 (22%) | 4.5 | 2.6–7.9 | 31.0 |
| +1355T/C[o] (C) | 53 (55%) | 43 (45%) | 131 (84%) | 25 (16%) | 4.2 | 2.3–7.5 | 25.0 |
| +1404 Fnu4Hl[o] (p) | 53 (55%) | 43 (45%) | 131 (84%) | 25 (16%) | 4.2 | 2.3–7.5 | 25.0 |
| +1428 Fokl[o] (a) | 53 (55%) | 43 (45%) | 131 (84%) | 25 (16% | 4.2 | 2.3–7.5 | 25.0 |
| +2331 A/T (A) | 48 (54%) | 41 (46%) | 68 (48%) | 75 (52%) | 0.8 | 0.5–1.3 | 0.9 |
| +2336 5bp del (ins) | 55 (57%) | 41 (43%) | 74 (47%) | 82 (53%) | 0.7 | 0.4–1.1 | 2.3 |
| +3201 Haell (p) | 82 (85%) | 14 (15%) | 127 (82%) | 27 (17%) | 0.8 | 0.4–1.6 | 0.4 |
| +3580 Mspl(p) | 42 (45%) | 52 (55%) | 49 (32%) | 106 (68%) | 0.6 | 0.3–1.0 | 4.3 |
| +3688 C/T (C) | 43 (45%) | 52 (55%) | 48 (31%) | 106 (69%) | 0.6 | 0.3–0.9 | 5.0 |
| +3839 A/wNl (p) | 95 (99%) | 1 (1%) | 153 (99%) | 1 (1%) | 1.6 | 0.2–11.6 | 0.1 |
| lGF2 Exon 3/Alul (a) | 34 (36%) | 60 (64%) | 41 (27%) | 113 (73%) | 0.6 | 0.4–1.1 | 2.5 |

The + allele is the most frequent allele in the diabetic population for diallelic polymorphisms, or the allele giving the maximum $\chi^2$ value for assocation. (From Lucassen et al. (1993), copyright ©by Nature Genetics and the authors. Reprinted with permission.)

a healthy person. The chi-square test based on (4.1.1) is given in the last column. Note that the $\chi^2$ values do not reveal the distance between the gene and a marker. A large sample size can result in a very large $\chi^2$ value with small association. The RR represents the **relative risk**, which is the same as odds ratio in most statistics books. It is defined as

$$RR = \frac{Pr\{\text{with diabetes} \mid ++\}/Pr\{\text{free from diabetes} \mid ++\}}{Pr\{\text{with diabetes} \mid +/- \text{ or } -/-\}/Pr\{\text{free from diabetes} \mid +/- \text{ or } -/-\}}$$
$$= \frac{p_{AB}p_{ab}}{p_{Ab}p_{aB}},$$

with the notation from Table 4.1.2 where A = diabetes, B = ++, a = normal, b = +/- or −/−. The column next to the right of RR is a 95% confidence interval for RR. A

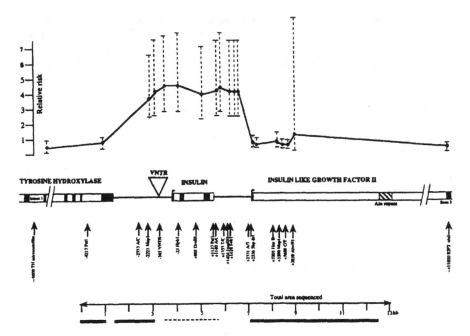

Figure 4.1.1: VNTR positions near INS area from Table 4.1.5. Positions of markers are shown with arrow signs. The graph shows the relative risk associated with each marker. Open boxes refer to introns; closed boxes to exons. The dashed lines correspond to the 95% C.L. in Table 4.1.5. (From Lucassen et al. (1993), copyright ©by Nature Genetics and the authors. Reprinted with permission.)

formula to compute this interval can be found in Agresti (1990, p.54). The formulae are as follows. Let RR be denoted by $\rho$. Using the observations in Table 4.1.2,

$$\hat{\rho} = \frac{(n_{AB} + 0.5)(n_{ab} + 0.5)}{(n_{Ab} + 0.5)(n_{aB} + 0.5)}.$$

It turns out that for large $n$,

$$\ln(\hat{\rho}) \sim N(\ln(\rho), \sigma^2) \quad \text{with} \quad \sigma^2 = \left(\frac{1}{n_{AB}} + \frac{1}{n_{aB}} + \frac{1}{n_{Ab}} + \frac{1}{n_{ab}}\right).$$

Fig. 4.1.1 and Table 4.1.5 apparently reveal that the disease gene must be between $-4127pst1$ and $+2331A/T$. They are only 7 kb apart, an impossible task for linkage analysis. (Remark: The authors of Lucassen et al. (1993) claim that the gene is between $-2733A/C$ and $+2331A/T$, a 4.1 kb interval.)

When there are more than two alleles in the marker locus, Table 4.1.1 has to be extended to an $m \times 2$ table. Small cells have to be combined in order to use the asymptotic $\chi^2$ test. Recent developments in this aspect can be found in Terwilliger (1995).

Instead of finding a control population, one may use the affected person's parents as controls. This is the transmission disequilibrium test (TDT) of Spielman et al. (1993). Suppose one affected individual has alleles (2, 4) and his parents have alleles (1, 2) and (3, 4), respectively. Apparently alleles 2 and 4 were transmitted to the affected child, but markers 1 and 3 were not. If many affected individuals are observed and the marker 2 is always transmitted to an affected child, we may suspect that this marker locus is closely linked to the disease gene and 2 is a particular allele that has not reached equilibrium with the disease gene.

Let T be the number of times a particular allele (in a given locus) that is transmitted to an affected child and R be the times that it is not. Under the null hypothesis that the marker and the disease gene has reached linkage equilibrium, we expect both $T \simeq (T + R)/2$ and $R \simeq (T + R)/2$, or the statistic,

$$Q_1 = \frac{(T - R)^2}{T + R} \tag{4.1.11}$$

should have an asymptotic $\chi_1^2$. If more than 1 allele are simultaneously considered, we may let $T_i$ be the number of times the $i$th allele is transmitted and $R_i$ be the number of times it is not transmitted, $i = 1, 2, \ldots, m$, where $m$ is the total number of alleles at this locus. Then an obvious extension of (4.1.11) is

$$Q_m = \sum_{i=1}^{m} \frac{(T_i - R_i)^2}{T_i + R_i}. \tag{4.1.12}$$

The distribution of $Q_m$ is less straightforward because the $T_i$'s and $R_i$'s are not independent, from $\Sigma T_i = \Sigma R_i$. Kaplan, et al. (1997) suggest a Monte-Carlo estimate for the distribution of $Q_m$, i.e., to simulate $Q_m$ under the null hypothesis that the disease gene and the markers are under linkage equilibrium. For example, if the parents have alleles $(i, j)$ and $(k, l)$, then their affected child should have an equal chance to be $(i, k)$, $(i, l)$, $(j, k)$ or $(j, l)$. A computer program is available from the authors via their an e-mail address given at the end of Kaplan et al.'s paper. They have also done a power study of this test and other tests. Please note that when (4.1.11) is used, data that provide no transmission information should be discarded, e.g., parents with marker types (1, 1) and (2, 2). This, however, will not affect the distribution by Monte-Carlo simulation. It is also apparent that under the null hypothesis of no linkage $\theta = 1/2$ (4.1.11) is valid regardless of the number of affected children per family. Thus, this method has great advantage over the affected-sibpair methods for diseases with low penetrance. To find families with two or more affected children is usually difficult.

To study how TDT is affected by alternative hypotheses, we need to find how (4.1.11), (4.1.12) and the disequilibrium coefficients are related. Let the disease allele be 1 and all the other alleles at this locus be 2 and let the penetrance of disease genotype $(ij)$ be $\lambda_{ij}$ with $\lambda_{11} \geq \lambda_{12} \geq \lambda_{22}$. Let the disequilibrium coefficient between disease allele $s(s = 1, 2)$ and marker allele $i(i = 1, 2, \ldots, m)$ be

$$d_{si} = p_{si} - q_s p_i,$$

where the notation follows (4.1.2) except the disease allele frequency is denoted by $q_s$ to avoid confusion. We wish to find the probability that $i$ is transmitted to an affected child but $j$ is not when one parent has marker genotype $(ij)(i \neq j)$. Let

| | | |
|---|---|---|
| $\theta$ | = | Recombination fraction between disease and marker loci |
| $MP$ | = | Genotype at marker locus of parent |
| $TM$ | = | Transmitted marker allele |
| $NM$ | = | Non-transmitted marker allele |
| $TD$ | = | Transmitted disease allele |
| $OTD$ | = | Disease allele transmitted by the other parent |
| $TH$ | = | Transmitted haplotype |
| $A$ | = | Affection status of offspring (1 = affected). |

Moreover, we assume $HWE$ for the disease gene alleles and the marker alleles. The probability we wish to find is

$$t_{ij} \equiv P\{TM = i, NM = j | A = 1\}.$$
$$= \frac{P\{TM = i, NM = j, A = 1\}}{P(A = 1)}. \tag{4.1.13}$$

Note that

$$P\{TM = i, NM = j, A = 1\}$$
$$= \sum_{s=1}^{2} P\{TH = si, NM = j, A = 1\}$$
$$= \sum_{s} P\{A = 1 | TH = si, NM = j\} P\{TH = si, NM = j\}$$
$$= \sum_{s} P\{A = 1 | TD = s\} P\{TH = si, NM = j\}. \tag{4.1.14}$$

The first term of (4.1.14) is

$$P(A = 1 | TD = s)$$
$$= \sum_{u=1}^{2} P(A = 1 | TD = s, OTD = u) P(OTD = u) = \lambda_{s1} q_1 + \lambda_{s2} q_2, \tag{4.1.15}$$

and the second term of (4.1.14) is

$$P(TH = si, NM = j) = p_{si} p_j (1 - \theta) + p_{sj} p_i \theta, \tag{4.1.16}$$

from the meiotic process. Putting (4.1.15)−(4.1.17) and

$$P(A = 1) = q_1^2 \lambda_{11} + 2q_1 q_2 \lambda_{12} + q_2^2 \lambda_{22},$$

into (4.1.13), we have

$$t_{ij} = p_i p_j \{1 + B[(1 - \theta) d_{1i}/(q_1 p_i) + \theta d_{1j}/(q_1 p_j)]\}, \tag{4.1.17}$$

where

$$B = [q_1^2(\lambda_{11} - \lambda_{12}) + q_{12}(\lambda_{12} - \lambda_{22})]/[P(A = 1)].$$

Thus, the probabilities of transmitting and non-transmitting $i$ to an affected offspring are respectively,

$$P(TM = i|A = 1) = \sum_{j=1}^{m} t_{ij} = p_i[1 + B(1 - \theta)d_{1i}/(q_1 p_i)]$$

$$\tag{4.1.18}$$

$$P(NM = i|A = 1) = \sum_{j=1}^{m} t_{ji} = p_i[1 + B\theta d_{1i}/(q_1 p_i)].$$

The above two probabilities evaluates the expected value of (4.1.11) and (4.1.12) under the non-null situation.

Equation (4.1.17) shows that in general $t_{ij} \neq t_{ji}$, but under $H_0 : \theta = 1/2$ or $d_{si} = 0$ for all $s$ and $i$, $t_{ij} = t_{ji} = 1/2$ for $i \neq j$. In other words $TDT$ is a test for both linkage and linkage disequilibrium, i.e. if $H_0$ is rejected, we accept both $\theta < 1/2$ and disequilibrium. Thus, $TDT$ is effective only when both conditions hold.

The test (4.1.12) consolidate the $t_{ij}$ into the marginal probabilities (4.1.18). We may test directly whether $t_{ij} = t_{ji}$ for all $i \neq j$ by (4.1.17) and (4.1.18). Sham and Curtis (1995), who derived (4.1.17), suggested two alternative models with three likelihood ratio tests. Let $n_{ij}$ be the number of parents who transmitted marker allele $i$ to an affected offspring and not transmitted allele $j$. Then the likelihood equation is

$$L \propto \prod_{i \neq j} (t_{ij})^{n_{ij}},\tag{4.1.19}$$

The null hypothesis and the two alternative models are:

$$H_0 : \theta = 1/2 \text{ or no disequilibrium } (t_{ij} = 1/2 \text{ for all } i, j),$$
$$H_1 : \ln (t_{ij}/t_{ji}) = b_i - b_j \text{ for all } t_{ij}( \text{ parsimonious model}),$$
$$H_2 : \text{General } t_{ij} \neq t_{ji}.$$

Sham and Curtis called $H_1$ versus $H_0$ the $G_1$ test, $H_2$ versus $H_0$ the $G_2$ test and $H_2$ versus $H_1$ the $G_{21}$ test. Obviously, $H_2$ is more general than $H_1$, but it requires more parameters. A computer program is available for all the three tests (see Appendix B.9). Note that $H_0$ includes the case $\theta < 1/2$ and all $d_{si} = 0$. In this case, the children's data are not independent under $H_0$ and (4.1.11) and (4.1.12) are valid only for one child per family. For extension to multi-children families, see Martin et al. (1997).

# 4.2   Gene, Exon and Intron Identification

The last stage in gene hunting is to sequence the gene and consequently to find the mutation that causes the disease. We have seen in the previous section that a statistical gene search method, whether it is by pedigree, sibpairs, or linkage disequilibrium, can confidently pinpoint a gene to a region around 1 cM. Since 1 cM contains roughly $10^6$ bp, it is unlikely that the whole region can be sequenced. But even if it is sequenced, we still do not know where the gene is because there may be more than one gene in that region. Moreover, a gene is usually represented by many exons separated by introns (see Fig. 1.3.4). For example, the cystic fibrosis gene discussed in Example 2.2.3 contains 27 exons with a total length of 4453 bp spanning about a 230 kb region at chromosome 7q31 (Harris et. al., 1996). One method to find the gene is first to find the cells that tend to use this gene to produce a particular protein. These cells are likely to contain the mRNA of this gene (see Fig. 1.3.5). Note that to have all these mRNAs alone is not sufficient to identify the gene because there may be hundreds of different mRNA simultaneously in existence at any given time. But if the gene is known to be in a 1 cM region and if we can identify one fragment of this region in the mRNA pool, then it is very likely this mRNA is responsible for the disease, because the chance that an mRNA is located in a 1 cM region is approximately 1/3000 (see Table 1.2.2.). Once the mRNA is identified, the composition of the gene is established and the mutation region can be found by comparing the mRNA in cells of a healthy and an affected individual.

To match a 1 cM region with many mRNAs is not an easy task. While it involves trial and error, the first thing we need is to make sure the DNA fragment sequenced in the gene candidate region belongs to an exon. Note that in the cystic fibrosis example just mentioned, the exon part is only 4453/230,000 or 2%. If an intron region is used to match the mRNAs, it is bound to fail. Of course, if the exon belongs to an irrelevant gene, there also would be no match. Many exon regions have to be tried, but we want to make sure that they are true exons. In addition to match mRNA from cells, the exons can also used to match protein segments or cDNAs in the gene bank. The name **cDNA** means complementary DNA which is the DNA complement of an existing RNA (Russell, 1990, Chapter 15, on how a cDNA library was formed).

The purpose of this section is to describe how to partition an uncharacterized DNA sequence into exon and intron regions. To identify exon-intron regions is also crucial for the application of the human genome project, which was 2% completed in 1997 (see Rowen et al., 1997 for details and information on other genomes, such as mouse and drosophila.). To be able to identify the exon, which is estimated to occupy only 5% of the human genome, will make its function easier to understand.

Exon identification has not yet reached very high accuracy. Much research still needs to be done. Two of the more recent methods are presented here.

## 4.2.1 Basic Ideas in Exon-Intron Identification

Fig. 1.3.5 showed how DNA signals are transcribed into mRNA. Note that the 5' to 3' non-transcription strand is the same as mRNA, except the $T$ in DNA is $U$ in mRNA. Thus, the 5' to 3' strand is usually used for exon-intron identification. Fig. 4.2.1 shows some basic information found at the intron-exon and exon-intron junctions. In this figure, the rectangular box represents an exon region. Its left end represents an intron-exon junction, also called the 3' splice site or acceptor splice site. It always ends with

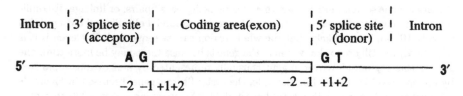

Figure 4.2.1: Transitions from intron to exon and exon to intron. The junction from intron to exon is a 3' splice site ending with an AG duplet, while the junction from exon to intron goes through a 5' splice site starting with GT. The coordinate system of the two junction starts with negative numbers on the left and positive numbers on the right. There is no site for the coordinate $\{0\}$.

the code-pair AG (with a few exceptions called nonconforming 3' site, see Senapathy et al., 1990). On the right side of the exon is the exon-intron junction, called the 5' splice site or donor splice site, which always starts with GT (again with a few exceptions called nonconforming 5' site, see again Senapathy et al., 1990). The splice sites are not coded for mRNA, and are deleted when the mRNA is formed (see again Fig. 1.3.5). The areas immediately adjacent to the splice sites usually have special structures that can be used to identify the junctions. However, unlike the AG and GT codes, the structures vary and can cause misidentification.

A typical exon region is also different from a typical intron region due to their different functions. However, this difference can only be recognized after a long walk in the region. An intuitive way to identify exon regions is to go through the following steps:

Step 1. Find all the AG and GT sites and consider all the regions flanked by them as exon candidates.

Step 2. Study the adjacent DNA structure of AG and GT sites and assess the likelihoods of their being junctions.

Step 3. Study the content between AG and GT and assess the likelihood of its being an exon.

Step 4. Combine the likelihoods of Step 2 and Step 3 and make an exon or intron classification on each block flanked by AG and GT.

We now turn to the question of how to assess the likelihoods mentioned in Steps 2 and 3. Suppose we wish to assess a DNA interval from a left position $L$ to a right position $R$, i.e., the information in $[L, R]$. Let the configuration in $[L, R]$ be $\psi$. For example, $\psi = TTACCG$ in $[+3, +8]$ at the 5' splice site using the coordinate system in Fig. 4.2.1. A commonly used likelihood assessment of $\psi$ being in an exon (symbol $E$) is the posterior distribution $P(E|\psi)$. If we let $I$ denote the event that $\psi$ belongs to an intron, then

$$P(E|\psi) = \frac{P(\psi|E)P(E)}{P(\psi|E)P(E) + P(\psi|I)P(I)}, \qquad (4.2.1)$$

where $P(E)$ and $P(I)$ are the prior probabilities of $\psi$ being from $E$ and $I$. The probabilities $P(\psi|E)$ and $P(\psi|I)$ are estimated from a sample, called the learning sample, from an existing genome. For example, based on 1333 primates, the donor-splice junction has probability 0.54, 0.02, 0.02, 0.42, respectively, for $\psi = A, T, C, G$ at position $+3$. These probabilities are quite different from $A, T, C, G$ at a typical intron site where they are approximately 0.36, 0.36, 0.14, 0.14, respectively (see Staden (1990)). However, these frequencies vary from species to species (see Klug and Cummings (1983, p. 148)), so large learning samples are necessary for exon identification.

When the window $[L, R]$ is small, it is possible to enumerate $P(\psi|E)$ and $P(\psi|I)$ for all the configurations. But even then, the tables are too numerous to list, they are stored in computers. For example, when $[L, R]$ contains 6 bp, called 6-tuple sequence, there are $4^6 = 4096$ frequencies (see Claverie et al. (1990)). When $[L, R]$ is longer, then it is difficult to effectively estimate all the configuration frequencies. Researchers usually use the average scores of many smaller partitions in $[L, R]$. The partitions may overlap each other depending on the information they can reveal. Averaged or not, the posterior probability is usually referred to as the **preference score**.

## 4.2.2 Linear Discriminant Analysis Approach

Basically, what we want to decide is whether a region flanked by AG and GT is an exon. Solovyev et al. (1994) used a seven variable linear discriminant function (see Appendix A.5.1) to accomplished this job. The 7 variables for the exon-intron (donor) junction discrimination are given in Fig. 4.2.2. They are

$x_1$ = average triplet preference scores in $[-30, -5]$

$x_2$ = average triplet preference scores in $[-4, +6]$ (conserved consensus region)

$x_3$ = average triplet preference scores in $[+7, +50]$ (G-rich region)

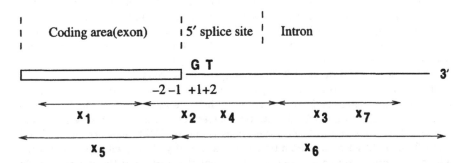

Figure 4.2.2: Positions of variables at the donor splice junction used in Solovyer et al. (1994). The lengths in the figure are not proportion to the number of base-pairs. The length for each variable is given in the text.

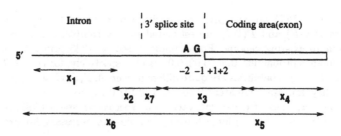

Figure 4.2.3: Positions of variables at the acceptor junction used in Solovyev et al. (1994). The descriptions are the same as in Fig. 4.2.2.

$x_4$ = number of significant triplets in $[-4, +6]$, (a significant triplet is defined as $P(E|\psi)$ in (4.2.1) $> 0.65$)

$x_5$ = average octanucleotide preference score in $[-60, -1]$

$x_6$ = average octanucleotide preference score of being intron in $[+1, +50]$

$x_7$ = the number of $G$-bases, $GG$-doublets and $GGG$-triplets in $[+6, +50]$

The prior distributions $P(E)$ and $P(I)$ in (4.2.1) are assumed to be equal when the preference scores are computed. The 7 variables for intron-exon (acceptor) junction discrimination are given as follows (see Fig. 4.2.3).

$x_1$ = average triplet preference score in $[-48, -34]$ (branch point region)

$x_2$ = average triplet preference score in $[-33, -7]$ (poly $(T/C)$-tract region)

$x_3$ = average triplet preference score in $[-6, +5]$ (conserved consensus region)

$x_4$ = average triplet preference score in $[+6, +30]$ (coding region)

$x_5$ = average octanucleotide preference score in $[+1, +54]$

$x_6$ = average octanucleotide preference score in $[-1, -54]$

$x_7$ = number of $T$ and $C$ in $[-33, -7]$ (poly $(T/C)$-tract region).

Those variables are chosen because of their effectiveness in discrimination. For example, the $x_7$ for intron-exon junction is included because in this region there tends to have a large number of $T$s and $C$s. From 1,333 primate intron-exon junctions examined, the $\Pr[T$ or $C]$ for any position in $[-4, -14]$ is between 0.75 and 0.88, much higher than their usual prevalence (Senapathy et al., 1990).

Solovyev et al. (1994) used 2037 donor splice sites, 2054 acceptor splice site and 89,417 pseudo-donor sites (containing a GT end but known as non-donor) and 134,150 pseudo-acceptor sites as learning sample. Their decision rule is the linear discrimination in the form of (A.5.8) with two classes. This procedure was used on a test set with 451 exons and 246,693 pseudoexons (nonexon DNA sequences flanked AG and GT) and obtained an exact internal exon sensitivity of 77% with specificity 79%, where an internal exon means an exon that is not the starting or the ending exon. In most cases, starting and ending exons are easy to recognize; a starting exon starts with ATG (because all protein starts initially with methionine, but it may be dropped later) and an ending exon ends with a stopping codon (see Table 1.3.1).

### 4.2.3 Neural Network Method: GRAIL II

GRAIL II (Xu et al., 1994) improved the old GRAIL (gene recognition and analysis internet link) that had been available for several years. Instead of using linear discrimination function, they used a neural network (see §A.5.2) for classification. The basic steps are shown in Fig. 4.2.4 and described as follows.

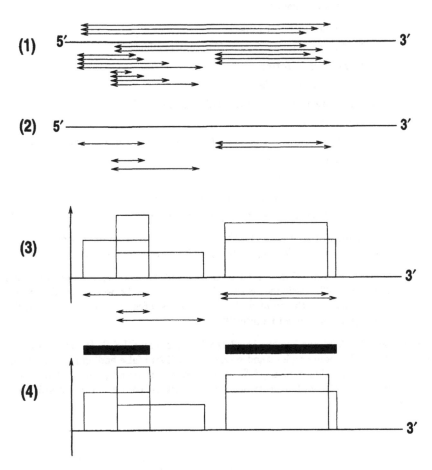

Figure 4.2.4: Basic steps in GRAIL II. (1) All blocks flanked by YAG and GT are considered as candidates. (2) Heuristic rules are used to remove most of the candidates in Step 1. (3) Neural networks are used to evaluate each block with a discriminant score, indicated by the heights of the rectangles in the figure. (4) Clustering method is used to summarize the results in Step 3. Selected exon regions are denoted by black bars.

Step 1. All the YAG and GT flanked blacks are considered as exon candidates, where Y represents one of the two pyrimidines, C or T.

Step 2. Thirty heuristic rules are used to rule out unlikely exon candidates. Approximately 95% of the candidates in Step 1 are eliminated. The rules are not given in the paper.

Step 3. Two hidden layers, one having 6 and one having 3 nodes were used. The neuron network is used to form a classification rule. Eleven variables, some of them were not clearly specified in their 1994 paper, are used as inputs for the neural network. They are:

$x_1$ = Preference score for 6-mers

$x_2$ = Preference score for 6-mers (frame-dependent)

$x_3$ = Markov chain fitting

$x_4$ = GC content of the isochore

$x_5$ = GC content of a 2 kb surrounding region

$x_6$ = Length preference score

$x_7$ = Acted length (normalized to [0.1] with 1 representing 300 bp or more)

$x_8$ = Donor junction score

$x_9$ = Acceptor junction score

$x_{10}$ = 6-mer in-frame score of 60 bp before the exon

$x_{11}$ = 6-mer in-frame score of 60 bp after the exon

The training sample had 110,000 bp. The output is an exon-score in [0, 1] indicating the candidate's possibility of being an exon (see (3) in Fig. 4.2.4).

Step 4. A clustering algorithm is used to merge and separate candidates with high exon-scores. The basic idea is to make sure that the between exon distance is not too small and the number of exons in a given region is not too large. Rules were again constructed empirically.

In a test set of 746 coding regions in 48 DNA sequences, GRAIL II recognized 697 (93.5%) of the 746 coding regions and predicted 93 (12%) false ending regions. In the correctly predicted coding regions, 62% of the exons were found perfectly (with both edges correct) and 93% with at least one edge correct. Apparently, there is still room for improvement.

# 4.3   Reconstruction of Evolutionary Path

The key parameter in evolutionary path (or **phylogenetic tree**) reconstruction is the time of separation between two species. It can be easily seen that if species $A$ and $C$ and

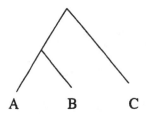

Figure 4.3.1: A simple phylogenetic tree.

$B$ and $C$ all have been separated for 10 million years and $A$ and $B$ have been separated for 5 million years, then we should have the evolutionary path shown in Fig. 4.3.1.

Besides using fossil records, the time of separation between species can be evaluated from the "similarity" between the existing species. The more similar the two species are, the shorter was the time since their separation. Though similarity can be measured by many means, such as anatomic structures, using genetic distance is one of the most objective ways to do it. In this section, two genetic distances, one based on allele frequency and one on DNA sequences, will be discussed.

### 4.3.1 Using Allele Frequency to Measure Genetic Distance

As we have seen in §1.2, the HWE equilibrium is reached after one generation under certain assumptions. One of the assumptions is that the population size should be large enough so that the real allele frequencies can be approximated by their expected values. This may not happen in nature because the population size is checked by availability of natural resources. This limitation on population size can force the allele frequency to change.

Let the population size be a constant $N$ in every generation and the frequency of allele $A$ be $p_0$ at the 0th generation. Then the frequency becomes

$$p_1 = \frac{X}{2N},$$

in the next generation, where X is a random variable representing the number of $A$ alleles in that generation. If we assume random mating, we have $X \sim Bi(2N, p_0)$ with mean and variance

$$Ep_1 = p_0, \quad \text{and} \quad \text{Var}[p_1] \equiv V_1 = \frac{p_0(1 - p_0)}{2N}.$$

Thus, in the 2nd generation, the mean of $p_2$ is still $p_0$, but its variance becomes

$$V_2 = E(p_2 - p_0)^2$$
$$= E_{p_1} E[((p_2 - p_1) + (p_1 - p_0))^2 | p_1]$$
$$= \frac{1}{2N} p_0(1 - p_0)\left(1 - \frac{1}{2N}\right) + \frac{p_0(1 - p_0)}{2N}.$$

In general, we have

$$V_{n+1} = \frac{p_0(1 - p_0)}{2N} + \left(1 - \frac{1}{2N}\right) V_n. \tag{4.3.1}$$

If we let $x_{n+1} = V_{n+1}/[p_0(1 - p_0)]$, (4.3.1) becomes

$$x_{n+1} = \frac{1}{2N} + \left(1 - \frac{1}{2N}\right) x_n, \tag{4.3.2}$$

which is a simple linear first order difference equation. Its solution is $x_n = 1 - (1 - 1/(2N))^n$, or

$$V_n \approx p_0(1 - p_0)(1 - e^{-n/(2N)})$$

$$\approx \frac{p_0(1 - p_0)}{2N} n.$$

Suppose each generation takes a constant time. Then the time of separation of two species is approximately

$$t \propto V_t/[p_0(1 - p_0)]. \tag{4.3.3}$$

There is no way to estimate $V_t$ from only one species, but with two species 1 and 2 from the same origin of the same population size, $V_t$ can be measured by the sample variance

$$\hat{V}_t = [(p_1 - \bar{p})^2 + (p_2 - \bar{p})^2/(2 - 1)$$

$$= (p_1 - p_2)^2/2,$$

where $p_1$ and $p_2$ are the two allele frequencies from species, 1 and 2, (note: the meaning of the subscripts in notation $p_i$ changes from generation to species). This is only for one allele. Let there be $k$ alleles, $A_1, A_2, \ldots, A_k$ at the same locus. Then a reasonable way to estimate the time of separation is

$$\hat{t} \propto \sum_{i=1}^{k} (p_{i1} - p_{i2})^2/[2\bar{p}_i(1 - \bar{p}_i)] \quad \text{with} \quad \bar{p}_i = (p_{i1} + p_{i2})/2,$$

where $p_{i1}$ and $p_{i2}$ denote the $i$th allele frequency for species 1 and 2, respectively. Usually, there is more than 1 locus we can measure. Suppose there are $L$ loci indexed by $\ell$. With an obvious increase in indexing, the final estimation of genetic distance between two species should be

$$\hat{t} \propto \frac{\sum_\ell \sum_i (p_{\ell i1} - p_{\ell i2})^2}{2 \sum_\ell \sum_i \bar{p}_{\ell i}(1 - \bar{p}_{\ell i})}. \tag{4.3.4}$$

The $\bar{p}$s are the averages of the $p$s from a sample. Formula (4.3.4), however, measures the relative time. To measure the real time, we need some estimate of $N$ and the time between each generation. The time of separation between two species can be measured in a different way if the mutation rate between alleles is known. This is Nei's standard genetic distance. Let the allele frequencies of two species 1 and 2 at time $t$ be

$$
\begin{array}{lcccc}
Z & 1 & 2 & \cdots\cdots & k \\
\text{Species} \quad 1 & x_1(t) & x_2(t) & \cdots\cdots & x_k(t) \\
\text{Species} \quad 2 & y_1(t) & y_2(t) & \cdots\cdots & y_k(t).
\end{array}
$$

At time 0, i.e., at the time of separation, we assume $x_i(0) = y_i(0)$ for all $i$. Suppose the allele frequency is in an equilibrium state. Suppose each year there is a probability of $\alpha$ that one allele is mutated to some new allele not listed in the table. Then it can be seen that the cross-product of the frequencies

$$
\sum x_i(t)y_i(t) \to \sum x_i(0)y_i(0)(1-\alpha)^{2t} \approx \sum x_i(0)y_i(0)e^{-2\alpha t} \tag{4.3.5}
$$

where $(1-\alpha)^{2t}$ represents the probability that the $i$th allele in the two species have not mutated. However, within the species, $\sum x_i(t)^2$ and $\sum y_i(t)^2$ tend not to change with time due to the equilibrium assumption. Thus, Nei's genetic identity defined by

$$
I(t) \equiv \frac{\sum_{\ell}\sum_{i} p_{\ell i 1}p_{\ell i 2}}{\sqrt{\sum_{\ell}\sum_{i} p_{\ell i 1}^2 \cdot \sum_{\ell}\sum_{i} p_{\ell i 2}^2}} \tag{4.3.6}
$$

tends to $(1-\alpha)^{2t} \sim e^{-2\alpha t}$, where all the notation follows from (4.3.4). Nei defined **maximum genetic distance** (Nei, 1987, p.221) as

$$
D(t) = -\ln I(t). \tag{4.3.7}
$$

From the current estimate $\alpha = 5 \times 10^{-7}/\text{year}$,

$$
\hat{t} = D/(2\alpha) = 10^6 D \text{ years.} \tag{4.3.8}
$$

Table 4.3.1 compares the divergence times by genetic distance (4.3.8) with other sources.

The distances (4.3.4) and (4.3.8) are only valid for short time periods of separation. (Note: short time in evolution means tens of thousands of years.) Otherwise the assumption of equilibrium and stable environment cannot hold even as a rough approximation. That is why in Table 4.3.1 the times of separation between species are relatively short. To quantify long time separation, we need to examine the DNA sequence itself.

## 4.3.2 Genetic Distance by DNA Sequence Substitution

DNA sequences that serve as basic life-sustaining protein codes are apparently derived from a common ancestor. Oxygen carriers and metabolism enzymes are some of the many examples. For a simpler model, Jukes and Cantor (1969) assume that the mutation rate of any nucleotide $A$, $T$, $C$ and $G$ to any other one be $\alpha t$ for a short time $t$. In other words, $P(A \to T) = P(A \to C) = P(A \to G) = \alpha t$, or the true mutation rate is $3\alpha$. Let

$$
q(t) = Pr\{\text{Two sites are still identical with a separation of time } t\}.
$$

Table 4.3.1: Estimated divergence times from genetic distance (in $10^5$ years) by (4.3.8) and by other sources.

| Organism | | Time estimate (years) from | | | |
|---|---|---|---|---|---|
| | $D$ | Distance $(\times 10^5)$ | Other sources $(\times 10^5)$ | Loci | $k$ $(\times 10^6)$ |
| Mammals | | | | | |
| Negroid & Mongoloid | 0.031 | $1.5 \pm 0,5$ | 1–2 | 62 | 5 |
| Man & chimpanzee | 0.62 | $40 \pm 10$ | 50 | 44 | (5) |
| Two macaque spp. | 0.11 | 5.4 | 4–5 | 28 | 5 |
| Pocker gophers | 0.08 | $1.2 \pm 0.8$ | 1–2 | 31 | 1.5 |
| Woodrat | 0.18 | $1.5 \pm 0.9$ | 2–4 | 20 | 0.8 |
| Deer mice spp. | 0.15 | $1.5 \pm 0.7$ | 1–5 | 28 | 1.0 |
| Ground squirrels (spp.) | 0.56 | $50 \pm 10$ | 50 | 37 | 6.7 |
| Ground squirrels | 0.10 | $6.9 \pm 0.3$ | 7 | 37 | 6.7 |
| Birds | | | | | |
| Galapagos finches | 0.12 | $6 \pm 3$ | 5–40 | 27 | 5 |
| Reptiles | | | | | |
| Bipes spp. | 0.62 | $31 \pm 10$ | 40 | 22 | 5 |
| Lizards | 0.28 | $50 \pm 25$ | c. 50 | 22 | 18 |
| Fishes | | | | | |
| Cave and surface fishes | 0.14 | $7 \pm 4.6$ | 3–20 | 17 | 5 |
| Minnows | 0.053 | 2.7 | 1–20 | 24 | 5 |
| Panamanian fishes | 0.32 | 58 | 20–50 | 28 | 18 |
| Ecninoids | | | | | |
| Panamanian sea urchins | 0.03–0.64 | – | 20–50 | 18 | |

*Notes:* spp. stands for subspecies, Loci = loci used for distance estimation, $k = 1/(2\alpha)$.
*Source:* From *Molecular Evolutionary Genetics* by Masatshi Nei, copyright ©1987 by Columbia University Press. Reprinted with permission of the publisher.

Thus,

$$q(t + \Delta t) = q(t)[P(\text{No mutation}) + P(\text{Both mutate to the same code})]$$
$$+ (1 - q(t))P(\text{Mutation occurred and the two sites become identical})$$
$$= q(t)[(1 - 3\alpha\Delta t)^2 + 3(\alpha\Delta t)^2] + (1 - q(t))[2(\alpha\Delta t)^2 + 2\alpha\Delta t(1 - 3\alpha\Delta t)],$$

where the second term means parallel mutation, i.e., two mutations to the same nucleotide, the third term also implies two mutations to the same nucleotide, but the last term indicates that one mutates to the other non-mutated one. Eliminating $(\Delta t)^2$ terms, we have

$$q(t + \Delta t) = (1 - 8\alpha\Delta t)q(t) + 2\alpha\Delta t.$$

Letting $\Delta t \to 0$, we have,

$$\frac{dq(t)}{dt} = -8\alpha q(t) + 2\alpha, \quad \text{with} \quad q(0) = 1.$$

The solution to the differential equation is

$$q(t) = 1 - \frac{3}{4}(1 - e^{-8\alpha t}),$$

which is equivalent to

$$t = \frac{-1}{8\alpha}\ln\left(1 - \frac{4}{3}(1 - q(t))\right). \tag{4.3.9}$$

Jukes and Cantor define their distance $K$ by the mean number of accumulated changes $2 \times 3\alpha t$, or

$$K = 6\alpha t = \frac{3}{4}\ln\left(\frac{3}{4q(t) - 1}\right). \tag{4.3.10}$$

Since $q(t)$ has a binomial distribution, and if $\hat{q}(t)$ is estimated based on $n$ base pairs, we have

$$\text{Var}(\hat{q}(t)) = q(t)(1 - q(t))/n$$

and

$$\text{Var}(\hat{K}) = \left(\frac{\partial K}{\partial q(t)}\right)^2 \text{Var}(\hat{q}(t)) = \left(\frac{3}{4q - 1}\right)^2 \text{Var}(\hat{q}(t)).$$

The DNA code contains two types of chemicals, purines $A$ and $G$ and pyrimidines $C$ and $T$. It turns out that it is easier to mutate between $A \rightleftharpoons G$ and $C \rightleftharpoons T$, than between $A$ to $T$ and $C$ and $G$ to $T$ and $C$ (see Fig. 4.3.2). The former is called

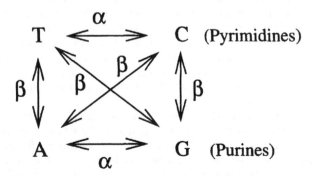

Figure 4.3.2: Two-parameter model of base substitution; $\alpha =$ rate of transition, $\beta =$ rate of transversion.

**transition-type substitution** and the later is called **transversion-type substitution**. Using the notation in Fig. 4.3.2, it has been found that in mitochondria DNA substitution, $\alpha = 18\beta$.

Let

$P_1(t) =$ Fraction of transitions
$P_2(t) =$ Fraction of transversions

We then have the following system of differential equations:

$$\frac{dP_1(t)}{dt} = 2\alpha - 4(\alpha + \beta)P_1(t) - 2(\alpha - \beta)P_2(t), \quad \text{and}$$

$$\frac{dP_2(t)}{dt} = 4\beta - 8\beta P_2(t), \quad \text{with} \quad P_1(0) = P_2(0) = 0.$$

The solution is

$$P_1(t) = \frac{1}{4}(1 - 2e^{-4(\alpha+\beta)t} + e^{-8\beta t})$$

$$P_2(t) = \frac{1}{2}(1 - e^{-8\beta t}).$$

This is called Kimura's two-parameter model (Kimura, 1980). The derivations are left as an exercise. The above equations are equivalent to

$$8\beta t = -\ln[1 - 2P_2(t)], \quad \text{and}$$

$$4\alpha t = -\ln[(1 - 2P_1(t) - P_2(t)] + \frac{1}{2}\ln[1 - 2P_2(t)].$$

The mean number of accumulated changes is

$$K \equiv 2(\alpha + 2\beta)t$$

$$= -\frac{1}{2}\ln[(1 - 2P_1(t) - P_2(t))\sqrt{1 - 2P_2(t)}]. \tag{4.3.11}$$

Weir (1996) compared the chicken and rabbit $\beta$-globin sequences. (The $\beta$-globin is one of the two parts in the hemoglobin that carries oxygen from the lungs to body cells.) The $K$ distances defined by (4.3.10) and (4.3.11) are respectively, $K = 0.3446$ (Jukes-Cantor's one-parameter model) and $K = 0.3513$ (Kimura's two-parameter model). The values are very similar. More detailed studies have found that even Fig. 4.3.2 can be improved. The four $\beta$'s may not be the same and the mutation rate from $T$ to $A$ may not be the same as that from $A$ to $T$. Some new results can be found in Lanave et al. (1984) and Yang (1994).

Since genome can be partitioned into intron and exon, it is natural to speculate that the mutation rates for the two parts may not be the same. There must be some restriction on mutations in exons, because these mutations may fatally change the protein structure. However, there are mutation in exons that do not change the protein composition. For example, codons $CUU$, $CUC$, $CUA$, $CUG$, $UUA$ and $UUG$ all represent the same animo acid leucine (see Table 1.3.1). Thus, the mutation from $CTT$ to $CTA$ in DNA will have no effect on the protein composition. The mutations that have the above property are called synonymous (silent) mutations, while a mutation that changes the amino acid a codon originally represents is called a nonsynonymous mutation. There are evidences that the synonymous mutations have a higher rate than the nonsynonymous mutations and the rate of synonymous mutations is similar to those in introns. Hughes and Yeager (1997) compared the substitution rates of 42 mouse and rat genes and found the average substitution rate for sites in introns was $0.201 \pm 0.008$(S.E.), for synonymous sites in exons was $0.197 \pm 0.014$ and for nonsynonymous sites in exons was $0.066 \pm 0.011$. Thus, it is better to estimate the separation of two species by synonymous substitution for short time separation and nonsynonymous substitution for long time separation. For a long duration (relative to the mutation rate), mutating back to the original nucleotide makes the true number of mutations difficult to estimate.

### 4.3.3 Reconstruction of Phylogenetic Tree

Given the distances between any two species, how do we construct a evolutionary tree that reflects these distance? Suppose the given distance matrix is shown as

$$D = \begin{bmatrix} 0 & d_{12} & d_{13} & \cdots & \cdots & d_{1n} \\ d_{21} & 0 & d_{23} & \cdots & \cdots & d_{2n} \\ \cdots & \cdots & \cdots & \cdots & \cdots & \cdots \\ d_{n1} & d_{n2} & d_{n3} & \cdots & \cdots & 0 \end{bmatrix},$$

where $d_{ij}$ denotes the distance between species $i$ and $j$. Obviously, the matrix is symmetric, i.e., $d_{ij} = d_{ji}$ for all $i$ and $j$. There are many methods of tree construction. It is in the area of cluster analysis. Here, we will present one of the most intuitive methods, the unweighted pair-group method using an arithmetic average (**UPGMA**). The steps can be most easily described by the following example. Table 4.3.2 gives the Jukes-Cantor $K$ distance (4.3.10) of five mitochondrial sequences of five primates.

Step 1. Link the two species with the closest distance. For Table 4.3.2 it is Human and Chimpanzee.

Step 2. Once the two species are linked, they become a new species and all the distances are adjusted as the average distance between the two species. For Table 4.3.2, human and chimpanzee are jointed as one new species called 1-2. The distance between 3 and 1-2 is the average distance between 3 and 1 and 3 and 2, i.e.,

$$d_{1-2,3} = (d_{31} + d_{32})/2 = (0.111 + 0.115)/2 = 0.113.$$

Now the grouping process goes back to step 1. After $n - 1$ groupings, the cluster analysis is done. In general, the distance between the combined species $A$ with members $\{a_1, a_2, \ldots, a_{n_1}\}$ and $B$ with members $\{b_1, b_2, \ldots, b_{n_2}\}$ is the average distance between all the pairs between $A$ and $B$.

The phylogenetic tree for Table 4.3.2 is shown in Fig. 4.3.3.

Table 4.3.2: Mitochondrial distances between primates.

|            | Human (1) | Chimpanzee (2) | Gorilla (3) | Orangutan (4) | Gibbon (5) |
|------------|-----------|----------------|-------------|---------------|------------|
| Human(1)      | – | 0.094 | 0.111 | 0.180 | 0.207 |
| Chimpanzee(2) | – | –     | 0.115 | 0.194 | 0.218 |
| Gorilla(3)    | – | –     | –     | 0.188 | 0.218 |
| Orangutan(4)  | – | –     | –     | –     | 0.216 |

A new distance matrix can now be constructed.

|                      | Human-Chimpanzee (1-2) | Gorilla (3) | Orangutan (4) | Gibbon (5) |
|----------------------|------------------------|-------------|---------------|------------|
| Human-Chimpanzee(1-2) | –  | 0.113 | 0.167 | 0.212 |
| Gorilla(3)            | –  | –     | 0.188 | 0.218 |
| Orangutan(4)          | –  | –     | –     | 0.216 |

*Source:* (From *Molecular Evolutionary Genetics* by Masatshi Nei, copyright ©1987 by Columbia University Press. Reprinted with permission of the publisher.)

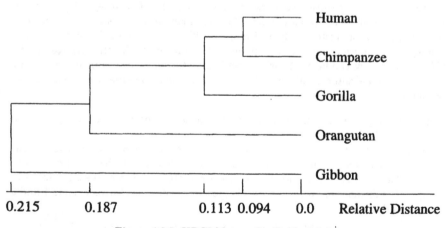

Figure 4.3.3: UPGMA tree for Table 4.3.2.

Fig. 4.3.4 gives one of the most recent evolutionary trees based on hemoglobin DNA using Jukes–Cantor distance and UPGMA clustering algorithm (by Czelusniak et al. in Doolittle, 1990, p. 611).

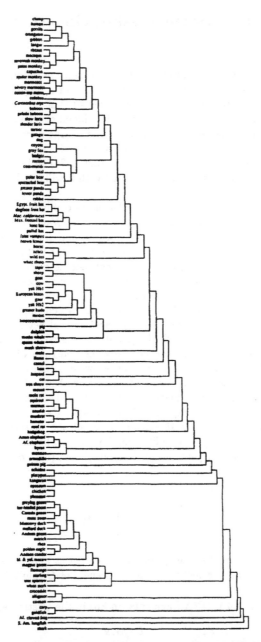

Figure 4.3.4: Phylogenetic tree by UPGMA clustering method based on 3034 nucleotide replacements on hemoglobin. (From Doolittle (1990), copyright © by Academic Press and the authors. Reprinted by permission.)

## 4.4   Forensic Evidence Using Genetic Markers

DNA markers, like finger prints, provide strong evidence in individual identification. They can be extracted from blood, semen, hair (root) or a fragment of skin that the criminal left at the crime scene. People have claimed that no two persons have the same finger prints and, except identical twins, this should also be true for DNA prints. However, one marker is not enough to differentiate a large number of persons. Many polymorphic markers are necessary for a positive identification. But how many markers are necessary and how polymorphic must they be? The basic statistical theory will be outlined and the the problems associated with it will also be introduced in this section.

In 1987, José Castro was accused of murdering his common-law wife Vilma Ponce and her two year old daughter. This was referred to as *New York vs Castro*, (Berry, 1991). Blood stains were found on Castro's watch and they were checked against Ponce's blood using four DNA probes, D2S44, D17S79, DXYS14 and DXY1. The last of these was used to ascertain gender and it will not be discussed in this example. Table 4.4.1 gives the match information of the first three probes. The numbers in the table are the DNA fragment lengths in kb.

Note that in this table D2S44 is a single band, which implies that the locus is homozygotic. The two bands of D17S79 represent typical heterozygotic alleles. The three bands of DXYS14 mean that the fragments associated with this probe appear in more than one locus in human chromosome. Now, *assume* these difference between the two blood samples are within measurement error and they can be considered as matches. What can we say about the chance of Castro being the murderer?

Let $m_1, m_2, \ldots, m_k$ be the $k$ markers to be examined. Let there be $\nu_i$ alleles, $a_{i1}, a_{i2}, \ldots, a_{i\nu_i}$ for marker $m_i$ with respective frequencies $f_{i1}, f_{i2}, \ldots, f_{i\nu_i}$. Assume that the genotype at each locus satisfies HWE and that all the markers are on different chromosomes without linkage disequilibrium. Then it can be shown that a randomly picked person with markers $\{(a_{11_1}, a_{11_2}), (a_{22_1}, a_{22_2}), \ldots, (a_{kk_1}, a_{kk_2})\}$ is

$$L_m = \prod_{i=1}^{k} g_i, \tag{4.4.1}$$

$$g_i = \begin{cases} 2f_{ii_1} f_{ii_2} & \text{if} \quad ii_1 \neq ii_2 \\ f_{ii_1}^2 & \text{if} \quad ii_1 = ii_2 \end{cases}$$

Let $X$ denote the marker configuration of a suspect and $Y$ denote the marker configuration the criminal evidence left at the scene. Let $G$ (guilty) denote the event

Table 4.4.1: DNA fragments (in kb) in the Castro case.

| Probe | D2S44 | D17S79 | | DXYS14 | | |
|-------|-------|--------|------|--------|-------|-------|
| Vilma Ponce | 10.162 | 3.869 | 3.464 | 4.855 | 2.999 | 1.946 |
| Blood on watch | 10.350 | 3.877 | 3.541 | 4.858 | 2.995 | 1.957 |

that a suspect is the criminal and $I$ (innocent) denote that he is not and $E$ be all the other evidence that is not related to the markers. Then what we wish to compute is the conditional probability

$$P(G|X,Y,E) \quad \text{or}$$
$$P(I|X,Y,E) = 1 - P(G|X,Y,E), \quad (4.4.2)$$

or, equivalently, the odds ratio

$$OD = \frac{P(G|X,Y,E)}{P(I|X,Y,E)}. \quad (4.4.3)$$

By $P(G|X,Y,E) = P(X,Y|G,E)P(G|E)P(E)/P(X,Y,E)$, (4.4.3) becomes

$$OD = \frac{P(X,Y|G,E)P(G|E)}{P(X,Y|I,E)P(I|E)}. \quad (4.4.4)$$

Since the evidence E is not related to the DNA evidence, (4.4.4) is further reduced to

$$OD = \left\{ \frac{P(X,Y|G)}{P(X,Y|I)} \right\} \left\{ \frac{P(G|E)}{P(I|E)} \right\}. \quad (4.4.5)$$

The first term on the right of Eq. (4.4.5) is the likelihood ratio and the second term is the prior distribution based on the other evidence E. Though the latter term is not related to genetic evidence, it is an indispensable part of the guilty claim. Those who are familiar with the application of the Bayes theorem know that (4.4.5) is the only correct way to compute (4.4.3). However, it is difficult to expect a juror in the court to come up with a number for $p(G|E)/p(I|E)$ (see comments by Kaye in Berry, 1991). We may totally discard all the other evidence and assume that

$$\frac{P(G|E)}{P(I|E)} = \frac{1}{N_0}, \quad (4.4.6)$$

where $N_0$ is all the possible persons who may have committed this crime. It can be all the adults in a city, a state, or in the whole world, though there should be no possibility that a native person in the Sahara Desert who knows no English would have committed a crime in New York city. This estimate is certainly in the suspect's favor, because any evidence will make (4.4.6) larger, except the suspect is identified solely by a DNA match in a police file. However, if the first term of (4.4.5) is very large, say $10^{15}$, then even the conservative estimate (4.4.6) of the second term will still provide a large odds ratio. Here is a quote from a DNA lab-test supervisor in the rape case *Martinez vs State of Minnesota* (see again comments by Kaye in Berry, 1991). The markers were taken from semen found at the crime scene and blood from the suspect.

Q: And what would be the answer to the question as far as the likelihood of finding another individual whose bands would match-up in the same fashion as this?

A: The final number was that you would expect to find one individual in 234 billion that would have the same banding that we found in this case.

Q: What is the total earth population, if you know?

A: Five billion.

Q: That is in excess of the number of the people today?
A: Yes. Basically that's what that number ultimately means, that the pattern is unique within the population of this planet.

The actual computation of the first term in (4.4.5) is very simple under the HEW and linkage equilibrium assumptions. For $X = Y$, a perfect match, $P(X, Y|G) = L_m$ and $P(X, Y|I) = L_m^2$ in (4.4.1). If a large number of highly polymorphic VNTRs (see §1.4) is used, a perfect match will be nearly impossible if the DNA of the suspect is not the one left at the scene. The only exception is when the suspect has an identical twin. Suppose 6 markers are used and each has 20 equally frequent alleles. Then under HWE, we have

$$\frac{P(X, Y|G)}{P(X, Y|I)} = \frac{1}{L_m} \approx \frac{1}{2^6 \times 20^{-12}} = 6.4 \times 10^{13}.$$

Note that the world population is $N_0 = 5 \times 10^9$.

Now, go back to the *New York vs Castro* example. If the defense lawyer looks into Table 4.4.1 carefully, he may raise the question whether the two bands of D2S44 were a match. Actually, according to Fig. 4.4.1, which shows the D2S44 distribution among the US Hispanic population to match Ponce's Hispanic origin, bands 10.162 and 10.350 can be two different alleles. Thus, the two bands can be claimed as mismatch, or a match within the limit of measurement error. In most forensic laboratories, a match is claimed if the difference between two bands is less than three standard deviations ($\sigma$) of the measurement. Experimental results show that the $\sigma$ is approximately $0.006 \times$ (*the fragment length*). If we consider the victim band is more accurately estimated due to the larger amount of blood available, we have $3 \times \sigma = 3 \times 0.006 \times 10.162 = 0.183$. The difference between the two estimates in Table 4.4.1 is $|10.162 - 10.350| = 0.188$, a little over $3\sigma$. The company that did the probing claimed it was a match. If the approximate $3\sigma$ rule is used, Figure 4.4.1 implies that the frequency of match is $4.9\% = 0.049$. Thus, the match of a homozygote band under HWE is $(0.049)^2 = 0.0024 = 1/417$. When the other likelihood of other matches are computed, the likelihood ratio against Castro was $725,000 : 1$. It was probably because of this overwhelming likelihood that Castro pleaded guilty.

First, we can see that there are several problems in using DNA evidence. This is why at the end of 1994 DNA evidence was not accepted in all states. The problems are (1) the HWE is not easy to establish, (2) the frequencies $f_{ij}$, which have to be estimated, vary in subpopulations (see Fig. 4.4.2), and (3) the identification of alleles is not without error. The three problems are not totally independent of each other. We will concentrate on the measurement error problem discussed by Berry (1991) and Berry et al. (1992). More opinions can be found in Weir (1992), Roeder (1994), Lander and Budowle (1994) and the book *DNA Technology in Forensic Science (1992)*.

We will first consider only one probe. Thus, the measured DNA fragment lengths $X$ and $Y$ in (4.4.3) are univariate random variables. Let the true lengths of $X$ and $Y$ be $\mu_x$ and $\mu_y$, and

$$X = \mu_x + \varepsilon_x, \quad \text{and} \quad Y = \mu_y + \varepsilon_y. \tag{4.4.7}$$

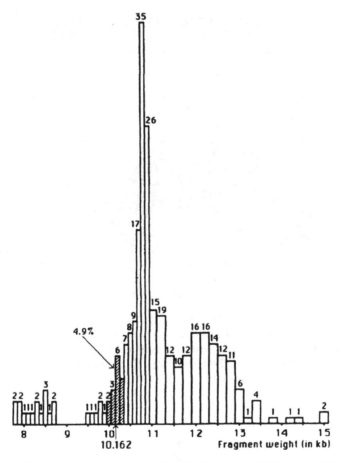

Figure 4.4.1: Frequency distribution of bands for probe D2S44 from 292 alleles in Hispanic population. (From Berry(1991), copyright © by the Institute of Mathematical Statistics. Reprinted by permission.)

Based on probing experience, it is reasonable to assume that both $\varepsilon_x$ and $\varepsilon_y$ are independent normal random variables with 0 mean and standard deviations $\sigma_x = c\mu_x$ and $\sigma_y = c\mu_y$ for some constant $c$. One such evidence is shown in Fig. 4.4.3, where the value

$$z = \frac{(x_1 - x_2)}{c(x_1 + x_2)/\sqrt{2}}, \tag{4.4.8}$$

is based on two repeated measures, $x_1$ and $x_2$, of the same fragment. Based on this evidence, the normality assumption for the $\varepsilon$s in (4.4.7) is quite reasonable.

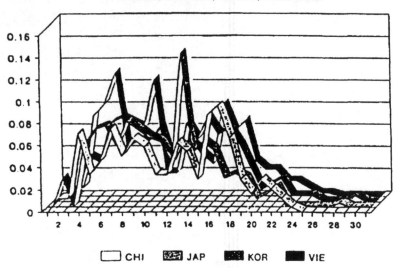

Figure 4.4.2: DNA profiles of $D10S28$ of four Asian subpopulations (Chinese, Japanese, Korean, and Vietnamese, from front to the back) in US. The horizontal axis is the bin number; bins are not of equal length. The sample sizes for the frequency estimation are around 150. (From Roeder(1994), copyright © by the Institute of Mathematical Statistics. Reprinted by permission.)

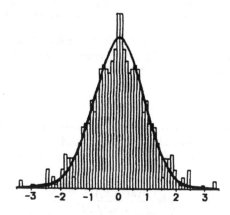

Figure 4.4.3: Histogram of (4.4.8) with $c = 0.008$ and 218 fragments. Note the symmetry is no surprise because both $x_1 - x_2$ and $x_2 - x_1$ are used, but the standard normal density function fits each side very well. Source: see Figure 4.4.4.

Let $G$ and $I$ denote the same events as the two in (4.4.3). Then

$$p(X = x, Y = y|G) = \int p(x, y|G, \mu_x = \mu_y = \mu) f(\mu) d\mu$$

$$= \int p(x, y|\mu_x = \mu_y = \mu) f(\mu) d\mu,$$

$$= \int p(x|\mu) p(y|\mu) f(\mu) d\mu, \tag{4.4.9}$$

and

$$p(X = x, Y = y|I) = p(x)p(y)$$

$$= \int_\mu p(x|\mu) f(\mu) d\mu \times \int_\mu p(y|\mu) f(\mu) d\mu, \tag{4.4.10}$$

where the density function $p(x|\mu)$ is

$$p(x|\mu) = \frac{1}{\sqrt{2\pi}c\mu} e^{-(x-\mu)^2/2c^2\mu^2} = \frac{1}{\sqrt{2\pi}c\mu} e^{-(x/\mu-1)^2/2c^2}, \tag{4.4.11}$$

and $f(\mu)$ is the density function of $\mu$, representing the population frequency distribution of $\mu$ such as in the histogram in Fig. 4.4.1. There is no unique way to estimate $f(\mu)$. It belongs to a large area in statistics called density estimation. Berry et al. (1992) suggest the following method.

Let $z_1, z_2, \ldots, z_n$ be the fragments from a sample of $n$ individuals. Berry et al. (1992) let

$$\hat{f}(\mu) = \frac{1}{n} \sum_{i=1}^n \frac{1}{\sqrt{2\pi}cb_n z_i} e^{-(\mu/z_i-1)^2/(2c^2b_n^2)}. \tag{4.4.12}$$

To see why this reasonable, suppose there is no experimental error. Then the sample $z_1, z_2, \ldots, z_n$ is equivalent to $\mu_1, \mu_2, \ldots, \mu_n$. A naive estimate of the distribution function is the empirical distribution based on the values $z_1, z_2, \ldots, z_n$. Many resampling schemes, such as the bootstrap method, use this approximation. But when error exists, the true $\mu_i$ is no longer $z_i$, but in the neighborhood of $z_i$. The neighborhood should be governed by

$$\frac{1}{\sqrt{2\pi}cb_n z_i} e^{-(\mu/z_i-1)^2/(2c^2b_n^2)}.$$

Putting equal weight on each $z_i$, we have (4.4.12). Since $c$ is a small number compared to $\mu$, if the sample size is small, the estimated frequency distribution could be unreasonably concentrated at a few locations. Thus, we wish $b_n$ to be large if $n$ is small and tends to 1 if $n$ is large. The optimal choice is unknown, but the smoothed curve is not sensitive to $b_n$ if $n$ is large. Putting (4.4.12) and (4.4.11) into (4.4.9) and (4.4.10), the first ratio in (4.4.5) is solved.

Berry et al. (1992) extended (4.4.7) to double bands at the same locus (see D17S79 in Table 4.4.1). In the case of a double band on one locus

$$\boldsymbol{X} = \begin{pmatrix} X_1 \\ X_2 \end{pmatrix}, \quad \text{and} \quad \boldsymbol{Y} = \begin{pmatrix} Y_1 \\ Y_2 \end{pmatrix}.$$

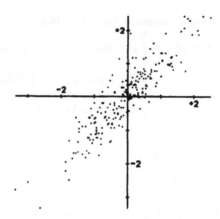

Figure 4.4.4: Scatterplot of 218 standard differences (4.4.14). $\hat{\rho} = 0.904$. (Both figures from Berry et al. (1992), copyright © by Blackwell Publishers Ltd. Reprinted by permission.)

We claim a match if $X = Y$ is within the measurement error. Assume there is no ambiguity between $X_1$ and $X_2$, i.e., the markers are chosen only when the two alleles are well separated in gel (and similarly for the $Y_1$ and $Y_2$). Then we may assume the following distributions: Let

$$
\begin{aligned}
X &\sim N\left(\begin{pmatrix} \mu_1 \\ \mu_2 \end{pmatrix}, c^2 \begin{bmatrix} \mu_1^2 & \rho\mu_1\mu_2 \\ \rho\mu_1\mu_2 & \mu_2^2 \end{bmatrix}\right), \\
Y &\sim N\left(\begin{pmatrix} \nu_1 \\ \nu_2 \end{pmatrix}, c^2 \begin{bmatrix} \nu_1^2 & \rho\nu_1\nu_2 \\ \rho\nu_1\nu_2 & \nu_2^2 \end{bmatrix}\right),
\end{aligned}
\tag{4.4.13}
$$

as an extension of (4.4.7), where the new notation should be self-explanatory. Since our interest is in the case when $X$ and $Y$ were from the same person, we consider only $\nu_1 = \mu_1$ and $\nu_2 = \mu_2$. Past data showed that $\rho$ may not be negligible. To estimate the magnitude of $\rho$, Fig. 4.4.4 shows the correlation coefficient between

$$
\left( \frac{(x_1 - y_1)}{c(x_1 + y_1)/\sqrt{2}}, \frac{(x_2 - y_2)}{c(x_2 + y_2)/\sqrt{2}} \right),
\tag{4.4.14}
$$

where $(x_1, y_1)$ and $(x_2, y_2)$ were two pair-bands of the same person, but evaluated at two different electrophoresis experiments. Since $(x_i + y_i)/2$ estimates $\mu_i$, $i = 1, 2$, the independence of $X$ and $Y$ implies

$$
\begin{aligned}
&Cov\left( \frac{(x_1 - y_1)}{c(x_1 + y_1)/\sqrt{2}}, \frac{(x_2 - y_2)}{c(x_2 + y_2)/\sqrt{2}} \right) \\
&\approx E \frac{[(x_1 - \mu_1) - (y_1 - \mu_1)][(x_2 - \mu_2) - (y_2 - \mu_2)]}{2c^2\mu_1\mu_2} = \rho.
\end{aligned}
$$

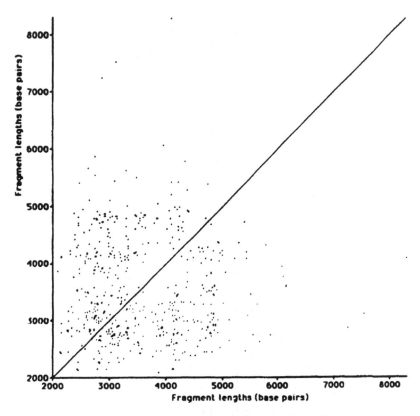

Figure 4.4.5: The relation between the two bands from one locus. Homozygotes are put in the diagonal. Based on 273 Caucasians using probe YNH24. (From Berry et al. (1992), copyright © by Blackwell Publishers Ltd. Reprinted by permission.)

When the correlation coefficient is estimated,

$$p(X = x, Y = y|G) = \iint p_{\mu_1\mu_2}(x_1, x_2)p_{\mu_1\mu_2}(y_1, y_2)f(\mu_1, \mu_2)d\mu_1 d\mu_2, \qquad (4.4.15)$$

and

$$p(X = x, Y = y|I)$$
$$= \iint p_{\mu_1\mu_2}(x_1, x_2)p_{\mu_1\mu_2}f(\mu_1, \mu_2)d\mu_1 d\mu_2 \iint p_{\mu_1\mu_2}(y_1, y_2)f(\mu_1, \mu_2)d\mu_1 d\mu_2, \qquad (4.4.16)$$

where $p_{\mu_1\mu_2}(\cdot, \cdot)$ is the bivariate density function for $X$, and $f(\mu_1, \mu_2)$ is the population distribution for the true lengths, $\mu_1$ and $\mu_2$. Berry et al. (1992) extended (4.4.12) to the bivariate distribution by letting

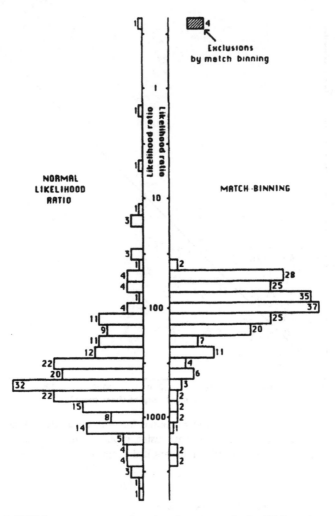

Figure 4.4.6: Within-person comparison by the two methods: NLR and matching-bin in finding the likelihood of two readingd from the same person when the readings did come from the same person. The method that gives larger values is the better method.

$$\hat{f}(\mu_1, \mu_2) = \frac{1}{n} \sum_{i=1}^{n} \frac{1}{2\pi c^2 b_n^2 z_{1i} z_{2i}} \exp\left\{ -\frac{1}{2c^2 b_n^2} \left[ \left( \frac{\mu_1}{z_{1i}} - 1 \right)^2 + \left( \frac{\mu_2}{z_{2i}} - 1 \right)^2 \right] \right\}, \quad (4.4.17)$$

with $b_n \geq 1$. Here $\mu_1$ and $\mu_2$ are assumed to obey HEW, i.e., they are independent. This is supported by data given in Fig. 4.4.5, where the pairs $(z_{1i}, z_{2i})$ are in random order. There is little correlation between $z_{1i}$ and $z_{2i}$ ($\hat{\rho} = -0.0005$). However, the

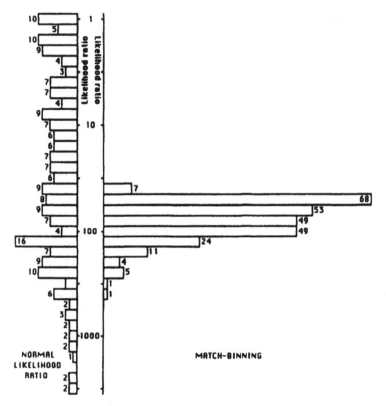

Figure 4.4.7: Between-person comparison of the two methods: NLR and matching-bin. In this case, the smaller the likelihood the better. (Both figures from Berry et al. (1992), copyright © by Blackwell Publishers Ltd. Reprinted with permission.)

HWE assumption needs to be checked when population changes. Extension to mutiple probes is straightfoward, but bands at different loci may not be independent due to disequilibrium in linkage. We will not pursue this subject further. The above procedure is called the **normal likelihood ratio method** (NLR). Berry et al. (1992) used the data of a 218 Caucasian (heterozygotes only) sample with duplicate measures on each individual and compared the NLR method with a standard match-binning method which claims match if the two lengths are within 2.475 standard deviations. The latter method cannot consider the correlation between the two bands. Fig. 4.4.6 shows the NLR method for matching two readings from the same person. In this case, we want the likelihood to be as large as possible. NLR is obviously the winner. On the other hand, there are $\binom{218}{2} = 23,653$ between person pairs in which we would like the likelihood of matching to be as small as possible. Fig. 4.4.7 again shows the the NLR method is better.

# Exercise 4

4.1 Forty three individuals were scored for two loci from a well mixed population, i.e., we may assume HWE at each loci. The following genotypes were found.

|     | AA | Aa | aa |
|-----|----|----|----|
| BB  | 16 | 9  | 2  |
| Bb  | 2  | 4  | 3  |
| bb  | 5  | 0  | 2  |

Is there evidence that there is a linkage disequilibrium between the two loci? (Hint: Estimate $p_{AB}$ by $(2 \times 16 + 2 + 9 + 4/4)/(2 \times 43)$.)

4.2 Suppose the current portions of haploid types $AB$, $Ab$, $aB$ and $ab$ are shown in Table 4.1.3. Let the linkage disequilibrium coefficient be $D$ and the recombination fraction between the two loci be $\theta, 0 < \theta < 1/2$. Assume the population is large and the mating is random. What is the $D$ in the next generation? Will $D$ converge to 0 as the number of generation increases?

4.3 Confirm the statistical results RR, 95% C.I. and $\chi^2$ in Table 4.1.5, using $-2733A/C(A)$ as an example.

4.4 Generalize the TDT test to QTL. Design the experiment, construct the test and describe how the power of your test can be computed. In computing the power, you need to specify the information you need.

4.5 Using DNA distance to infer the age of separation depends on the molecular evolutionary clock hypothesis, i.e., there is a constant mutation rate. How do we test the validity of this hypothesis? Consider the tree in Fig. 4.3.1. Let an aligned DNA code at position $P$ for the three species be $P_A, P_B$, and $P_C$ respectively. Let

$$n_{ijk} = \text{number of sites has } (P_A, P_B, P_C) = (i, j, k)$$

where $i, j, k = A, T, C,$ or $G$. We are interested in testing whether $H_0$: the mutation rate in the paths to A and to B are the same. How do we do the test? (Hint: under $H_0 : En_{jik} = En_{jik}$; Tajima, 1993).

4.6 In (4.3.1) we have the difference equation,

$$V_{n+1} = \frac{p_0(1 - p_0)}{2N} + \left(1 - \frac{1}{2N}\right) V_n.$$

Solve this difference equation.

4.7 Suppose the cDNAs found at certain stage of the embryonic development are concentrated on a few chromosomes. Then we may trace how genes are arranged in the chromosomes according to the developmental order. Suppose an animal has $k$ chromosomes and suppose that in the embryonic cells, out of $n$ randomly typed mRNAs, $r$ of them are on the same $m$ chromosomes. Is this events unusual? How do you test the statistical significance of your claim? Suppose $m$ is a specified small number a priori. (see application in Tchernev et al., 1997).

4.8  Hartmann et al. (1994) examined the diversity of races in US based on VNTR RFLP at 4 loci. Nei's genetic distances were computed in the following table. Use UPGMA to cluster the past evaluation tree of the 6 races. (Note this data is based on 4 loci only. It is subject to some bias.)

| /   | Chinese | Japanese | Korean | Vietnamese | Black | White | Hispanic |
|-----|---------|----------|--------|------------|-------|-------|----------|
| C   | –       | 0.024    | 0.004  | 0.021      | 0.040 | 0.047 | 0.032    |
| J   | ***     | –        | 0.012  | 0.024      | 0.020 | 0.037 | 0.019    |
| K   | ***     | ***      | –      | 0.015      | 0.026 | 0.027 | 0.023    |
| V   | ***     | ***      | ***    | –          | 0.034 | 0.043 | 0.018    |
| B   | ***     | ***      | ***    | ***        | –     | 0.023 | 0.021    |
| W   | ***     | ***      | ***    | ***        | ***   | –     | 0.010    |
| H   | ***     | ***      | ***    | ***        | ***   | ***   | –        |

4.9  Use Table 1.3.1 to compute the proportion of silent mutations in all the amino acid codons.

4.10  We have derived that the odds in (4.4.4) $P(X, Y|G)/P(X, Y|I) = 1/L_m$. However, $f_{ij}$s in (4.4.1) have to be estimated? What will the defendant's lawyer say? What will the prosecuting lawyer say? Use the 10.162 bin in Fig. 4.4.1 to construct a 95% upper bound for its frequency.

4.11  The value of VNTR locus for DNA fingerprint depends on its probability for false matching, i.e., the possibility of two unrelated DNA are falsely matched. Suppose a VNTR locus has $n$ alleles with frequencies $p_1, p_2, \ldots, p_n$. What is the chance that two randomly picked persons will have a perfect match? What is the optimum distribution of the $p$'s that will make a false match minimum? Suppose we can use one locus with $n = 2m$ alleles of equal frequency or two loci each with $m$ equal frequency alleles. Which choice is better from false matching viewpoint?

4.12  A gene has two alleles $A$ and $a$. We wish to estimate the population frequency $p_A$ of $A$. Let $n$ individuals be randomly selected and their genotypes be typed. There are $n_1$ $AA$ type, $n_2$ $Aa$ type and $n_3$ $aa$ type $(n_1 + n_2 + n_3 = n)$. An obvious estimate for $p_A$ is

$$\hat{p}_A = \frac{2n_1 + n_2}{2n}$$

Show that, under HWE, $\hat{p}_A$ is unbiased with variance $p_A(1 - p_A)/(2n)$, but its variance becomes

$$V(\hat{p}_A) = \frac{p_A + p_{AA} - 2p_A^2}{2n},$$

if we cannot assume HWE, where $p_{AA}$ is the population frequency for $AA$.

4.13 Following data is a small portion of the data used to find a diabetes gene using marker A/G-AT (with 18 alleles). The full data file, *Ex2 − 13.dat*, is in Yang's web site *AppB*.9. Use the *ETDT* software to check whether there is linkage between this locus and diabetes.

| Family | Father's marker type | Mother's marker type | Affected child's marker type |
|--------|----------------------|----------------------|------------------------------|
| D1-NM2 | (12, 12) | (1, 13) | (12, 13) |
| D1-NM2 | (12, 12) | (1, 13) | (12, 1) |
| D1-NM3 | (1, 16) | (1, 12) | (1, 12) |
| D1-NM3 | (1, 16) | (1, 12) | (1, 12) |
| D4 | (1, 1) | (1, 17) | (1, 1) |
| D4 | (1, 1) | (1, 17) | (1, 1) |
| ... | ... | ... | ... |

# Appendix A

# Statistical Results Useful in Genetics

This appendix serves as a reminder of some statistical results that are useful in genetics. Most formulae are listed without derivation. They may be difficult to follow without sufficient background in statistics. Capital letters usually represent random variables and small letters represent data. Bold-faced letters are usually vectors or matrices.

## A.1  Multivariate Analysis and Regression

Let

$$X = \begin{bmatrix} X_1 \\ X_2 \\ \vdots \\ X_p \end{bmatrix}$$

be a vector of $p$-variate normal random variables with the mean

$$\mu = \begin{bmatrix} \mu_1 \\ \mu_2 \\ \vdots \\ \mu_p \end{bmatrix} \quad \text{and} \quad \text{covariance matrix } \Sigma = (\sigma_{ij}) = (Cov(X_i, X_j)).$$

Its density function is

$$f(x) = ((2\pi)^p |\Sigma|)^{-\frac{1}{2}} e^{-\frac{1}{2}(x-\mu)' \Sigma^{-1}(x-\mu)}. \tag{A.1.1}$$

The usual notation is $X \sim N(\mu, \Sigma)$. If $A$ is an $m \times p$ matrix of constants, then $Y \equiv AX \sim N(A\mu, A\Sigma A')$. This can be shown by the expansion of $Y_i = \sum a_{ij} X_j$, where $Y_i$, $a_{ij}$ are elements in $Y$ and $A$. The correlation coefficient $\rho_{ij}$ between $X_i$ and $X_j$ is defined as

$$\rho_{ij} = \frac{\sigma_{ij}}{\sigma_i \sigma_j}, \quad \text{where} \quad \sigma_i = \sqrt{\sigma_{ii}}, \; \sigma_j = \sqrt{\sigma_{jj}}.$$

If the $X$ vector is partitioned into two parts as

$$X = \begin{bmatrix} X_a \\ X_b \end{bmatrix}, \tag{A.1.2}$$

and their means and covariance matrix are also accordingly partitioned as

$$\mu = \begin{bmatrix} \mu_a \\ \mu_b \end{bmatrix}, \quad \Sigma = \begin{bmatrix} \Sigma_{aa} & \Sigma_{ab} \\ \Sigma_{ba} & \Sigma_{bb} \end{bmatrix},$$

then the conditional distribution of $X_a$ given $X_b = x_b$ is

$$X_a | x_b \sim N(\mu_a + \Sigma_{ab} \Sigma_{bb}^{-1}(x_b - \mu_b), \quad \Sigma_{aa} - \Sigma_{ab} \Sigma_{bb}^{-1} \Sigma_{ba}). \tag{A.1.3}$$

In particular, for $p = 2$, $X_a = X_1$ and $X_b = X_2$

$$X_1 | x_2 \sim N(\mu_1 + \frac{\sigma_1}{\sigma_2} \rho_{12}(x_2 - \mu_2), \quad \sigma_1^2(1 - \rho_{12}^2)). \tag{A.1.4}$$

The covariance matrix $\Sigma_{a \cdot b} \equiv \Sigma_{aa} - \Sigma_{ab} \Sigma_{bb}^{-1} \Sigma_{ba}$ in (A.1.3) gives the conditional covariance of elements in $X_a$ conditioned on $X_b$. If we let the $(i, j)$th entry of $\Sigma_{a \cdot b}$ be $\sigma_{ij \cdot b}$, then

$$\rho_{ij \cdot b} \equiv \frac{\sigma_{ij \cdot b}}{\sqrt{\sigma_{ii \cdot b} \sigma_{jj \cdot b}}}$$

is called the partial correlation coefficient between $X_i$ and $X_j$ conditional on $X_b$. In particular, if $X_b$ contains only one element, $X_3$, then

$$\Sigma_{a \cdot b} = \begin{pmatrix} \sigma_{11} & \sigma_{12} \\ \sigma_{21} & \sigma_{22} \end{pmatrix} - \begin{pmatrix} \sigma_{13} \\ \sigma_{33} \end{pmatrix} \sigma_{33}^{-1} (\sigma_{13} \; \sigma_{23}),$$

or

$$\rho_{12 \cdot 3} = \frac{\sigma_{12} - \sigma_{13} \sigma_{33}^{-1} \sigma_{32}}{\{(\sigma_{11} - \sigma_{33}^{-1} \sigma_{13}^2)(\sigma_{22} - \sigma_{33}^{-1} \sigma_{23}^2)\}^{1/2}}. \tag{A.1.5}$$

Let the data gathered for $X$ be $x_1, x_2, \ldots, x_n$. Then by the definition of the mean and covariance matrix, $\mu$ should be estimated by

$$\hat{\mu} = \bar{x} = \sum_{h=1}^{n} x_h / n,$$

and $\Sigma$ by

$$\hat{\Sigma} = \sum_{h=1}^{n} (x_h - \bar{x})(x_h - \bar{x})' / (n - 1). \tag{A.1.6}$$

Let the $ij$th entry in $\hat{\Sigma}$ be $\hat{\sigma}_{ij}$. Then the sample correlation coefficient, called the Pearson correlation coefficient $r_{ij}$, defined as

$$r_{ij} = \frac{\hat{\sigma}_{ij}}{\{\hat{\sigma}_{ii}\hat{\sigma}_{jj}\}^{1/2}},$$

is used as an estimate of $\rho_{ij}$. The distribution of $r_{ij}$, under the assumption that the true correlation $\rho_{ij} = 0$ is

$$r_{ij}\sqrt{n-2} \Big/ \sqrt{1-r_{ij}^2} \sim t_{n-2}. \tag{A.1.7}$$

(Anderson, 1958, p. 65). For large $n$,

$$r_{ij} \sim N(0, 1/(n-3)).$$

A good approximation of the distribution of $r_{ij}$ for large $n$ ($n > 25$) with true $\rho_{ij}$ not necessarily 0 is

$$\sqrt{n-3}(z-\eta) \sim N(0, 1), \tag{A.1.8}$$

with

$$z = \frac{1}{2}\ln\frac{1+r_{ij}}{1-r_{ij}}, \quad \eta = \frac{1}{2}\ln\frac{1+\rho_{ij}}{1-\rho_{ij}}, \tag{A.1.9}$$

(see Anderson, 1958, p. 77). This is called "Fisher's $z$-transform." Distribution (A.1.8) can be used to construct confidence intervals for $\rho_{ij}$. The distribution of Fisher's $z$ holds also for the partial correlation coefficient $\rho_{ij\cdot b}$, i.e., when $\rho$ and $r$ are replaced by $\rho_{ij\cdot b}$ and $r_{ij\cdot b}$ in (A.1.9), the distribution (A.1.8) still holds, except the term $\sqrt{n-3}$ should be replaced by $\sqrt{n-|b|-3}$, where $|b|$ denotes the number of elements in $X_b$ (see Anderson, 1958, p. 86). The term $r_{ij\cdot b} \equiv \hat{\rho}_{ij\cdot b}$ is computed in the obvious way by replacing the terms in $\Sigma_{ij\cdot b}$ with their estimates following (A.1.6) and (A.1.3).

The SAS PROC CANCORR can be used for correlation and partial correlation computations.

One can see from the definition of a correlation that it is invariant under linear transformation, i.e. the correlation between $X_i$ and $X_j$ is the same as that between $(X_i - \mu_i)/\sigma_i$ and $(X_j - \mu_j)/\sigma_j$ for any constants $\mu_i, \mu_j, \sigma_i \neq 0$ and $\sigma_j \neq 0$. In particular, if we let the $\mu$s and $\sigma$s be their corresponding means and standard deviations, then $(X_i - \mu_i)/\sigma_i$ and $(X_j - \mu_j)/\sigma_j$ both have a standard normal distribution and their covariance is the same as their correlation. In most practical situations, correlation, instead of covariance, is used to quantify the relation between two random variables because the former is unit invariant. Thus, for the rest of this section, we let all the $X_i$ be standardized, i.e., they all have 0 mean and variance 1. In particular, (A.1.5) becomes

$$\rho_{12\cdot 3} = \frac{\rho_{12} - \rho_{13}\rho_{32}}{\sqrt{(1-\rho_{13}^2)(1-\rho_{23}^2)}}.$$

It is rather intuitive that the regression equation, which states the relation between one variable $X_1$ and other variables $x_2, \ldots, x_p$ must be closely related to the correlation coefficients between them. If we re-examine (A.1.3), we see that $X_a$ can be best predicted by $\mu_a + \Sigma_{ab}\Sigma_{bb}^{-1}(x_b - \mu_b) = \Sigma_{ab}\Sigma_{bb}^{-1}x_b$, where $\mu$s are omitted because all the $X$s are standardized. In other words, the best prediction (in terms of prediction error) for $X_1$ given $x_b = [x_2, x_3, \cdots, x_p]'$, is

$$x_1 = \Sigma_{1b}\Sigma_{bb}^{-1}x_b$$
$$\equiv \beta_2 x_2 + \beta_3 x_3 + \cdots + \beta_p x_p, \qquad\qquad \text{(A.1.10)}$$

where the $\beta$s are defined by the expansion of $\Sigma_{1b}\Sigma_{bb}^{-1}$. It can be seen that the $\beta$s in (A.1.10) should be the regression coefficient. If we write (A.1.10) in the usual regression equation (no constant term due to standardization) with $n$ observations as

$$x_{h1} = x_{h2}\beta_2 + x_{h3}\beta_3 + \cdots + x_{hp}\beta_p + \varepsilon_h, \quad h = 1, 2, \ldots, n, \qquad \text{(A.1.11)}$$

or in the matrix form

$$Y = X\beta + \varepsilon,$$

where $Y = [x_{11}, x_{21}, \ldots, x_{n1}]'$, $X$ is a $n \times (p-1)$ matrix with $(i, j)$th entry $x_{ij}$, and $\varepsilon$ is the error vector, then

$$\hat{\beta} = [X'X]^{-1}[X'Y] = [(X'X)/n]^{-1}[X'Y]/n. \qquad \text{(A.1.12)}$$

The true $\beta$ is the $\hat{\beta}$ when $n \to \infty$, or $\beta = \Sigma_{bb}^{-1}\Sigma_{b1}$ as defined by (A.1.10). Thus, in genetics, as well as in many other applications, the regression coefficients are also called the *partial* regression coefficients. For example, $\beta_2$ quantifies the relationship between $x_1$ and $x_2$ when all the other variables $x_3, x_4, \ldots, x_p$ are fixed. To make the notation clearer, $\beta_j$ is also denoted as $\beta_{1j \cdot b(j)}$, where $b(j)$ is the set $[2, 3, \ldots, j-1, j+1, \ldots, p]$. There is a theoretical difference between (A.1.10) and (A.1.11). Since (A.1.10) is based on multivariate analysis, $x_1$ and $x_2, \ldots, x_p$ should be observed simultaneously, but in (A.1.11), $x_{h2}, \ldots, x_{hp}$ are supposed to have been given first as the design variables and then $x_{h1}$ is *produced*. A typical example for the former is weight and height and the latter is fertilizer level and yield. However, in many practical situations, they are considered exchangeable, e.g., we may try to find a regression equation between weight and height or to find the correlation between fertilizer level and yield. In genetics, no matter whether the variables are genotypes or phenotypes, they are usually multivariate data. However, regression equations are often used for easy interpretation and computation. For a theoretical discussion, the reader may refer to Sampson (1974) or Stuart and Ord (1991, Vol. 2, Chapter 27). Let $c$ represent the set $\{3, 4, \ldots, p\}$. Though the regression coefficient $\beta_{12 \cdot c}$ and $\rho_{12 \cdot c}$ have similar interpretations, there is no one-to-one correspondence between them. The reason is simple, $\rho_{12 \cdot c}$ and $\rho_{21 \cdot c}$ have to be the same but $\beta_{12 \cdot c}$ does not necessarily have to be equal to $\beta_{21 \cdot c}$. To express $\beta_{12 \cdot c}$ in terms of the correlations, we may use the relationship between (A.1.10) and (A.1.11),

$$(\beta_2, \beta_3, \ldots, \beta_p) = \Sigma_{1b}\,\Sigma_{bb}^{-1}$$
$$= [\sigma_{12}\,\Sigma_{1c}]\begin{bmatrix} \sigma_{22} & \Sigma_{2c} \\ \Sigma_{c2} & \Sigma_{cc} \end{bmatrix}^{-1}. \qquad \text{(A.1.13)}$$

Applying the well-known formula

$$\begin{bmatrix} A & B \\ C & D \end{bmatrix}^{-1} = \begin{bmatrix} [A - BD^{-1}C]^{-1} & -A^{-1}B[D - CA^{-1}B]^{-1} \\ -D^{-1}C[A - BD^{-1}C]^{-1} & [D - CA^{-1}B]^{-1} \end{bmatrix}$$

to (A.1.13), we have

$$\beta_2 = \beta_{12\cdot c} = \sigma_{12}(\sigma_{22} - \Sigma_{2c}\Sigma_{cc}^{-1}\Sigma_{c2})^{-1} - \Sigma_{2c}\Sigma_{cc}^{-1}\Sigma_{c2}(\sigma_{22} - \Sigma_{2c}\Sigma_{cc}^{-1}\Sigma_{c2})^{-1}$$

$$= \frac{\sigma_{12} - \Sigma_{1c}\Sigma_{cc}^{-1}\Sigma_{2c}}{\sigma_{22} - \Sigma_{2c}\Sigma_{cc}^{-1}\Sigma_{c2}} = \frac{\sigma_{12\cdot c}}{\sigma_{22\cdot c}}.$$

Similarly, we have $\beta_{21\cdot c} = \sigma_{12\cdot c}/\sigma_{11\cdot c}$, or

$$\rho_{12\cdot c}^2 = \beta_{12\cdot c} \cdot \beta_{21\cdot c}. \tag{A.1.14}$$

This relationship appears also in simple linear regression where $\rho_{xy}^2 = \beta_{y|x} \cdot \beta_{x|y}$ and $\beta_{y|x}$ being the regression coefficient for the model $y = \alpha + \beta x + \varepsilon$ and $\beta_{x|y}$ for $x = \alpha + \beta y + \varepsilon$. One important indicator of a regression model is the rate of variance reduction $R^2$,

$$R^2 \equiv 1 - \hat{\varepsilon}'\hat{\varepsilon}/Y'Y = 1 - (Y - X\hat{\beta})'(Y - X\hat{\beta})/(Y'Y)$$

$$= YX'(X'X)^{-1}X'Y/(Y'Y) \tag{A.1.15}$$

$$\to \Sigma_{1b}\Sigma_{bb}^{-1}\Sigma_{b1},$$

where the 1 in $\Sigma_{1b}$ indicates the elements $x_1$ and $b = \{2, 3, \ldots, p\}$. Note in the derivation of (A.1.15), we used the fact that $(Y'Y)/n \to I$ as $n \to \infty$, because all the variables are standardized. In multivariate analysis, $R = \sqrt{R^2}$ is called the **multiple correction coefficient** between $X_1$ and $X_2, X_3, \ldots, X_p$. It is the maximum correlation coefficient between $X_1$ and any linear combination of $X_2, X_3, \ldots, X_p$.

# A.2  Likelihood Ratio Test and General Linear Models

Let the likelihood function of $x = \{x_1, x_2, \ldots, x_n\}$ for parameter $\theta = (\theta_r, \theta_s)$ be

$$L(x|\theta) = \prod_{i=1}^{n} f(x_i|\theta), \tag{A.2.1}$$

and the two hypotheses under testing be

$$H_0 : \theta_r = \theta_{r0}$$
$$H_1 : \theta_r \neq \theta_{r0}. \tag{A.2.2}$$

Note that under $H_1$, $\boldsymbol{\theta}_r$ should be flexible enough to take values in a set $\Omega$. The likelihood ratio is defined as

$$\lambda = \frac{L(x|\boldsymbol{\theta}_{r0}, \tilde{\boldsymbol{\theta}}_s)}{L(x|\hat{\boldsymbol{\theta}}_r, \hat{\boldsymbol{\theta}}_s)}, \tag{A.2.3}$$

where $\tilde{\boldsymbol{\theta}}_s$ is the maximum likelihood estimate (mle) of $\boldsymbol{\theta}_s$ of (A.2.1) under $H_0$ and $(\hat{\boldsymbol{\theta}}_r, \hat{\boldsymbol{\theta}}_s)$ are the mle of $(\boldsymbol{\theta}_r, \boldsymbol{\theta}_s)$ under $H_1$.

If $\boldsymbol{\theta}_{r0}$ is an interior point of $\Omega$, then for large $n$, we have

$$\begin{aligned} -2\ln\lambda &\sim \chi_r^2(\delta^2) \quad \text{with } \delta^2 = 0 \text{ under } H_0, \quad \text{and} \\ \delta^2 &= (\boldsymbol{\theta}_r - \boldsymbol{\theta}_{r0})'\vartheta^{-1}(\boldsymbol{\theta}_r - \boldsymbol{\theta}_{r0}), \quad \text{under } H_1, \end{aligned} \tag{A.2.4}$$

where $\chi_r^2(\delta^2)$ is the chi-square distribution with $r$ degrees of freedom with non-centrality parameter $\delta^2$, and $\vartheta$ is the information matrix with its the $ij$'s entry being

$$\omega_{ij} = -E\left[\frac{\partial^2 \ln L}{\partial\theta_i \partial\theta_j}\right], \quad \text{with} \quad \theta_i, \theta_j \in \boldsymbol{\theta}_{r0}.$$

Thus, the power of the likelihood test at $\alpha$ level is

$$\text{Power} = Pr\{\chi_r^2(\delta^2) > \chi_{r,\alpha}^2\}, \tag{A.2.5}$$

where $\chi_{r,\alpha}^2$ is the upper $\alpha 100\%$ level of a chi-square distribution with $r$ degrees of freedom. (Stuart and Ord, 1991, Vol. II, pp. 861, 869).

In case $\boldsymbol{\theta}_{r0}$ is not an interior point of $\Omega$, such as $\boldsymbol{\theta}_{r0}$ is on the boundary of $\Omega$, then the asymptotical result $-2\ln(\lambda) \sim \chi_r^2$ is not true under $H_0$. This is similar to one-sided and two-sided options in the usual $t$ or $Z$ test. The $p$-value for the one-sided test is only half of that for the two-sided test. In genetics, if the test is on a recombination fraction with $H_0 : \theta = 1/2$ versus $H_1 : \theta < 1/2$, the likelihood ratio statistic $-2\ln\lambda$ tends to a $50:50$ $\{0\}$ and $\chi_1^2$, i.e.,

$$Pr\{-2\ln\lambda > x_0\} = 0.5Pr\{\chi_1^2 > x_0\} \tag{A.2.6}$$

for any threshold $x_0 > 0$. Results for more complex boundaries can be found in Self and Liang (1987) and Holmans (1993).

Note: The LOD score is

$$LOD = -\log_{10}\lambda = -0.4343\ln\lambda = 0.2171(-2\ln\lambda).$$

**Example A.2.1**  *In Stuart and Ord (1991, II), where (A.2.5) was derived, the following example is provided: Let the data $x_1, x_2, \ldots, x_n$ be $N(\mu, \sigma^2)$ with both parameters being unknown. The purpose is to test $H_0 : \sigma^2 = \sigma_0^2$ versus $H_1 : \sigma^2 \neq \sigma_0^2$. It can be shown that*

$$\lambda = \left(\frac{s^2}{\sigma_0^2}\right)^{n/2} \exp\left[-\frac{1}{2}n\left\{\frac{s^2}{\sigma_0^2} - 1\right\}\right].$$

It can be seen that

$$\delta^2 = -E\left[\frac{\partial^2 \ln L}{\partial(\sigma^2)^2}\right](\sigma^2 - \sigma_0^2)^2 = \frac{n}{2\sigma_0^4}(\sigma^2 - \sigma_0^2)^2.$$

Let $n = 48$ and the true $\sigma^2 = 1.5\sigma_0^2$. Then $\delta^2 = 6.0$ and the power by (A.2.5) at $\alpha = 0.05$, by the asymptotic test $-2\ln\lambda \to \chi_1^2$, is

$$\text{Power} = Pr\{\chi_r^2(\delta^2) > \chi_{r,0.05}^2\} = Pr\{\chi_1^2(6.0) > 3.84\} = 0.688.$$

The SAS output (pw1) is given following Example A.2.2.    □

The theory of a general linear model and analysis of variance can be derived from the principle of the likelihood ratio test. In a general linear model

$$Y = X\theta + \varepsilon \equiv (X_1\ X_2)\begin{pmatrix}\theta_r \\ \theta_s\end{pmatrix} + \varepsilon, \text{ with } \varepsilon \sim N(0, \sigma^2 I),$$

where

$$X_1(n \times r \text{ matrix}) = \begin{bmatrix} x_{11}' \\ x_{21}' \\ \vdots \\ x_{n1}' \end{bmatrix}, \quad X_2(n \times s \text{ matrix}) = \begin{bmatrix} x_{12}' \\ x_{22}' \\ \vdots \\ x_{n2}' \end{bmatrix},$$

and the two hypotheses are the same as (A.2.2). The likelihood ratio becomes

$$\lambda = \frac{\max_{\theta_s,\sigma^2}(2\pi\sigma^2)^{-n/2}\exp\{-\sum_i^n(y_i - x_{i1}'\theta_{r0} - x_{i2}'\theta_s)^2/(2\sigma^2)\}}{\max_{\theta_r,\theta_s,\sigma^2}(2\pi\sigma^2)^{-n/2}\exp\{-\sum_i^n(y_i - x_{i1}'\theta_r - x_{i2}'\theta_s)^2/(2\sigma^2)\}}$$

$$= \frac{(2\pi\hat{\sigma}_1^2)^{-n/2}e^{-n/2}}{(2\pi\hat{\sigma}_0^2)^{-n/2}e^{-n/2}} \propto \hat{\sigma}_1^2/\hat{\sigma}_0^2, \text{ where}$$

$$\hat{\sigma}_1^2 = (Y - X_1\theta_{r0})'(I - X_2(X_2'X_2)^{-1}X_2')(Y - X_1\theta_{r0})/n \equiv \text{SSE}_1/n$$

$$\hat{\sigma}_0^2 = Y'(I - X(X'X)^{-1}X')Y/n \equiv \text{SSE}_0/n,$$

where $\text{SSE}_1$ and $\text{SSE}_0$ are the two commonly used notations for error sums of squares under $H_1$ and $H_0$, respectively.

If we let $k = r + s$ and the sample size be $n$, then the likelihood ratio test is equivalent to

$$F = \frac{(SSE_0 - SSE_1)/r}{SSE_1/(n - k)} \sim F_{n-k}^r \tag{A.2.7}$$

under $H_0$, where $F$ is the F-distribution with $(r, n - k)$ degrees of freedom. Under $H_1$, if the $\theta_r$ is a fixed effect, then (A.2.7) has F-distribution with $(r, n - k)$ degrees of freedom with noncentrality parameter

$$\delta^2 = (E[SSE_0 - SSE_1] - rE[MSE])/E[MSE]$$
$$= r(E[MSR] - E[MSE])/E[MSE], \tag{A.2.8}$$

where MSR is the mean square due to regression, defined as $(SSE_0 - SSE_1)/r$, and MSE is the mean square due to error (residuals), defined as $SSE_1/(n - k)$. (A.2.8) is based on the fact that the numerator of (A.2.7) has either a central or noncentral chi-square distribution and we know that

$$E[\chi^2_r(\delta^2)] = r + \delta^2,\qquad\qquad\text{(A.2.9)}$$

and $(SSE_0 - SSE_1)/\sigma^2$ has a chi-square distribution and $\sigma^2 = E[MSE]$. This leads to (A.2.8). Depending on the design, the expression of the expected mean squares may not be in a simple form. It is the EMS column in most analysis of variance tables. If we use the formula for the expectation of a quadratic form

$$E(\boldsymbol{Y'AY}) = \boldsymbol{\mu'_y A\mu_y} + \text{tr}\boldsymbol{A\Sigma_y}, \quad \text{with} \quad \boldsymbol{\mu_y} = E[\boldsymbol{Y}], \ Cov(\boldsymbol{Y}) = \boldsymbol{\Sigma_y},$$

we see that, under $H_1$,

$$E[SSE_0] = (n - k)\sigma^2, \quad \text{by} \quad \boldsymbol{\Sigma_y} = \sigma^2 I, \quad \text{and} \quad \text{tr}(\boldsymbol{I} - \boldsymbol{X}(\boldsymbol{X'X})^{-1}\boldsymbol{X'}) = n - k$$
$$E[SSE_1] = (\boldsymbol{\theta_r} - \boldsymbol{\theta_{r0}})'\boldsymbol{X'_1}(\boldsymbol{I} - \boldsymbol{X_2}(\boldsymbol{X'_2X_2})^{-1}\boldsymbol{X'_2})\boldsymbol{X_1}(\boldsymbol{\theta_r} - \boldsymbol{\theta_{r0}}) + (n - s)\sigma^2.$$

When $\boldsymbol{\theta_r}$ are random effects, we usually assume that $\boldsymbol{\theta_r} \sim N(0, \sigma^2_r I)$, independent of $\boldsymbol{\varepsilon}$. In this case, $E[SSE_0] = \sigma^2(n - k)$ still holds, but

$$E[SSE_1] = \text{tr}(\boldsymbol{I} - \boldsymbol{X_2}(\boldsymbol{X'_2X_2})^{-1}\boldsymbol{X'_2})(\sigma^2 I + \sigma^2_r\boldsymbol{X_1X'_1})$$
$$= (n - s)\sigma^2 + \sigma^2_r\text{tr}\boldsymbol{X'_1}(\boldsymbol{I} - \boldsymbol{X_2}(\boldsymbol{X'_2X_2})^{-1}\boldsymbol{X'_2})\boldsymbol{X_1}.$$

The numerator of (A.2.7) now becomes a central chi-square distribution when normalized by the variance E[MSR]. (Note: noncentrality comes from the non-zero mean part of the normal distribution, but $\theta_r$ has mean 0.) Hence (A.2.7) becomes a weighted central F distribution. Let us use a balanced two-way mixed model as an example. Let the row with $a$ levels be a fixed affect, the column with $b$ levels be a random effect, and there be $n$ observations per each column-row combination. Then the ANOVA table can be written as Table A.2.1 (Montgomery, 1984, p. 220).

The notation is standard and they will not be defined here. If we apply (A.2.8) to test the row (fixed) effect, we have

$$\frac{SSA/(a - 1)}{SSAB/((a - 1)(b - 1))} \sim F^{a-1}_{(a-1)(b-1)}(\delta^2) \quad \text{with} \quad \delta^2 = bn \sum_{i=1}^{a}(\alpha_i - \bar{\alpha})^2/(\sigma^2 + n\sigma^2_{\alpha\beta}).$$

$$\text{(A.2.10)}$$

Thus the power of the test at $\alpha$ level is

$$\text{Power} = Pr\{F^{a-1}_{(a-1)(b-1)}(\delta^2) > F^{a-1}_{(a-1)(b-1),\alpha}\}.\qquad\text{(A.2.11)}$$

If we apply (A.2.8) when we test the column (random) effect, we have

$$\frac{SSB/(b - 1)}{SSE/(ab(n - 1))}\frac{\sigma^2}{(\sigma^2 + an\sigma^2_{\beta})} \sim F^{b-1}_{ab(n-1)}.$$

Table A.2.1 Analysis of variance table for two way mixed model.

| Source | S.S. | D.F. | EMS |
|--------|------|------|-----|
| Row (A, $\alpha$) | SSA | $a-1$ | $\sigma^2 + n\sigma_{\alpha\beta}^2 + bn\sum_{i=1}^{a}(\alpha_i - \bar{\alpha})^2/(a-1)$ |
| Column (B, $\beta$) | SSB | $b-1$ | $\sigma^2 + an\sigma_{\beta}^2$ |
| Interaction | SSAB | $(a-1)(b-1)$ | $\sigma^2 + n\sigma_{\alpha\beta}^2$ |
| Error | SSE | $ab(n-1)$ | $\sigma^2$ |

Hence the power of the the test at $\alpha$ level is

$$\text{Power} = Pr\left\{ F_{ab(n-1)}^{b-1} > \frac{\sigma^2}{(\sigma^2 + an\sigma_\beta^2)} F_{ab(n-1),\alpha}^{b-1} \right\}. \tag{A.2.12}$$

To make sure we have the right functions to compute these powers and consequently determine the sample sizes, the following numerical examples are used for illustration.

**Example A.2.2**   In Table A.2.1, let $a = 3$, $b = 4$, $n = 2$, $\sigma^2 = 1$, $\sigma_\beta^2 = 2$, $\sigma_{\alpha\beta}^2 = 0.5$ and $\sum_{i=1}^{3}(\alpha_i - \bar{\alpha})^2 = 2.5$. Then the power to test the fixed effect is, by (A.2.11),

$$\text{Power} = Pr\{F_6^2(\delta^2) > F_{6,0.05}^2\} = Pr\{F_6^2(10) > 5.14\} = 0.58.$$

And to test the random effect, the power is

$$\text{Power} = Pr\left\{ F_{12}^3 > \frac{1}{1+3\cdot2\cdot2} F_{12,0.05}^3 \right\} = Pr\{F_{12}^3 > 0.268\} = 0.847. \qquad \square$$

The F-values are shown in the following SAS output and they have been confirmed by the operation characteristic curves in Montgomery (1984).

```
- - - - - - - - - - - - - - - SAS Input - - - - - - - - - - - -
PROC IML;

/*    For Example A.2.1            */

w=3.84;
pwo-1-probchi(w,1,0);
pw1=1-probchi(w,1,6.0);
print pw0[format=5.3] pw1[format=5.3];

/*    For Example A.2.2            */

x=5.14;
px=1-probf(x,2,6,10);
y=0.268;
py=1-probf(y,3,12,0);
```

```
print x px[format=5.3] y py[format=5.3];
FINISH;
RUN;
- - - - - - - - - - - - - - Output - - - - - - - - - - - - - -
         PW0    PW1
        0.050  0.688

         X     PX          Y     PY
        5.14  0.580       0.268  0.847
```

## A.3 Multinomial Distribution

The theory of multinomial distribution plays an important role in genetics because many estimations are based on the allele counts which have a multinomial distribution. Let there be $k$ classes with true proportions $\{p_1, p_2, \ldots, p_k\}$ with $\sum p_i = 1$. Let the observed counts in the $i$th class be $n_i$ subject to a fixed number of total counts $n$. Then $\{n_1, n_2, \ldots, n_k\}$ has a multinomial distribution with

$$L(n_1, n_2, \ldots, n_k | p_1, p_2, \ldots, p_k) = n! \prod_{i=1}^{k} \frac{p_i^{n_i}}{n_i!}. \tag{A.3.1}$$

Its moment generating function is

$$\phi(t) \equiv E\left[\prod_{i=1}^{k} e^{n_j t_j}\right] = \left(p_1 e^{t_1} + p_2 e^{t_2} + \cdots p_k e^{t_k}\right)^n. \tag{A.3.2}$$

(A.3.2) is very useful for computing the moments for the multinomial distribution. It can be shown (Johnson and Kotz, 1969) that

$$E[n_i] = np_i, \quad \mathrm{Var}[n_i] = np_i(1 - p_i), \quad \text{and} \quad \mathrm{Cov}(n_i, n_j) = -np_i p_j. \tag{A.3.3}$$

For moments, it is easier to find them through cumulants. The formula higher can be found in Johnson and Kotz (1969, p. 284) and Stuart and Ord (1987, Vol I, pp. 86, 87, 105, and 196).

Suppose the probabilities $\{p_1, p_2, \ldots, p_k\}$ depend on one unknown parameters $\theta$. Then the mle estimate of $\theta$ based on (A.3.1) is asymptotically unbiased with variance $\vartheta^{-1}$, where $\vartheta$ is the Fisher's information

$$\vartheta = -E \frac{\partial^2 \ln L}{\partial \theta^2} = n \sum_{i=1}^{k} \frac{1}{p_i} \left(\frac{\partial p_i}{\partial \theta}\right)^2. \tag{A.3.4}$$

**Example A.3.1** *(from Elandt-Johnson, 1971, p. 296). Suppose there are two alleles, A and a, in a locus and the aa has a disadvantage in fetal development. Thus the $F_2$ intercross ratio of $AA : Aa : aa$ is no longer $1 : 2 : 1$, but $1 : 2 : \phi$. In an experiment using 200 intercrosses of Aa, the observed frequencies were $AA : Aa : aa = 58 : 129 : 13$. Estimate $\phi$ and the estimation accuracy.*

**Sol:** After normalizing the ratios, the probabilities $P_i$'s are $1/(3 + \phi)$, $2/(3 + \phi)$ and $\phi/(3 + \phi)$. The mle for $\phi$ is to solve

$$\frac{\partial \ln L}{\partial \phi} = \frac{n_3}{\phi} - \frac{n}{3 + \phi} = 0.$$

Note that

$$\frac{\partial^2 \ln L}{\partial \phi^2} = -\frac{n_3}{\phi^2} + \frac{n}{(3 + \phi)^2}.$$

It can be seen that

$$\mathrm{Var}(\hat{\phi}) \approx \frac{\phi(3 + \phi)^2}{3n}.$$

Substituting the data, we have $\hat{\phi} = 0.2086$ and $\hat{\sigma}_{\phi}^2 = 0.00358$, or $\hat{\sigma}_{\hat{\phi}} = 0.06$.  □

When the probabilities $\{p_1(\theta), p_2(\theta), \ldots, p_k(\theta)\}$ depends on a set of unknown parameters $\boldsymbol{\theta} = \{\theta_1, \theta_2, \ldots, \theta_s\}$, the information matrix is still manageable. Let

$$u_j(\ell, \boldsymbol{\theta}) = \frac{1}{p_\ell} \frac{\partial p_\ell}{\partial \theta_j}.$$

Then

$$(\vartheta)_{ij} = n \sum_{\ell=1}^{k} p_\ell u_i(\ell, \boldsymbol{\theta}) u_j(\ell, \boldsymbol{\theta}). \tag{A.3.5}$$

The mle is usually harder to obtain than the moment estimates. Bailey's theorem is very useful in estimating parameters in discrete data.

**Bailey's Theorem** *Suppose there are s parameters to be estimated in a discrete data set of k categories $(k > s)$. Suppose the total number of observations is fixed, and further restrictions (reasonable, see the proof) make the data have s degrees of freedom. Then the moment estimate is equivalent to the mle.*

PROOF. Let the observed frequency of the $i$th category be $n_i$ and the true probability for this category be $p_i(\boldsymbol{\theta})$. Then the method of moments is to solve any $s$ independent equations. Without loss of generality, let them be

$$n_i = np_i(\boldsymbol{\theta}), \quad i = 1, 2, \ldots, s. \tag{A.3.6}$$

Let the restriction for all the other $n_i$'s be

$$n_i = f_i(n_1, n_2, \ldots, n_s), \quad \text{for} \quad i = s + 1, \ldots, k. \tag{A.3.7}$$

Since (A.3.7) has to be true for all $\{n_i\}$'s, it is still true if we double these $n_i$, triple them, etc. It can be shown that if $f$ is continuous,

$$p_i(\boldsymbol{\theta}) = f_i(p_1(\boldsymbol{\theta}), p_2(\boldsymbol{\theta}), \ldots, p_s(\boldsymbol{\theta})), \quad \text{for} \quad i = s+1, \ldots, k. \qquad (A.3.8)$$

Consequently,

$$\begin{aligned} np_i(\boldsymbol{\theta}) &= f_i(np_1(\boldsymbol{\theta}), np_2(\boldsymbol{\theta}), \ldots, np_s(\boldsymbol{\theta})) \\ &= f_i(n_1, n_2, \ldots, n_s), \ (\text{by}(A.3.6)) \\ &= n_i. \ (\text{by} \ (A.3.7)). \end{aligned}$$

Combining with (A.3.6), we have

$$np_i(\boldsymbol{\theta}) = n_i, \quad \text{for all} \quad i = 1, 2, \ldots, k. \qquad (A.3.9)$$

The mle of (A.3.1) requires

$$\frac{\partial \ln L}{\partial \theta_j} = \sum_{i=1}^{k} \frac{n_i}{p_i(\boldsymbol{\theta})} \frac{\partial p_i(\boldsymbol{\theta})}{\partial \theta_j} = 0. \qquad (A.3.10)$$

If we solve (A.3.6) or, equivalently, (A.3.9), (A.3.10) becomes

$$n \sum_{i=1}^{k} \frac{\partial p_i(\boldsymbol{\theta})}{\partial \theta_j},$$

which is 0, as required by mle. Note that $\sum_{i=1}^{k} p_i(\boldsymbol{\theta}) = 1$. □

# A.4   Inference on Frequency Tables

(i) For a frequency table (histogram):

In a frequency table (histogram), let $o_1, o_2, \ldots, o_k$, be observed in categories, $1, 2, \ldots, k$ respectively. Let $p_1, p_2, \ldots, p_k$ be the expected proportions in these categories under the null hypothesis $H_0$, $\sum_{j=1}^{k} p_j = 1$. Then the p-value to reject $H_0$ and to accept the alternative hypothesis, $H_1$ that $H_0$ is not true, is

$$p\text{-value} = Pr\left\{ \chi_\nu^2 > \sum_{j=1}^{k} \frac{(np_j - o_j)^2}{np_j} \right\},$$

where $n = \sum_{j=1}^{k} o_j$, and $\nu = k - 1 - m$, with $m$ = number of parameters that have to be estimated from the data in order to calculate the $p_j$s.

(ii) To test for independence (association) between two traits.

Let the counts $o_{..}$ be observed as the following $r \times c$ contingency table.

| | Column | | | | $r$ |
|---|---|---|---|---|---|
| | 1 | 2 | $\cdots$ | $c$ | Total |
| | $o_{11}$ | $o_{22}$ | $\cdots$ | $o_{1c}$ | $n_1.$ |
| | $o_{21}$ | $o_{22}$ | $\cdots$ | $o_{2c}$ | $n_2.$ |
| | $\cdots$ | $\cdots$ | $\cdots$ | $\cdots$ | $\cdots$ |
| Row | $\cdots$ | $\cdots$ | $\cdots$ | $\cdots$ | $\cdots$ |
| | $\cdots$ | $\cdots$ | $\cdots$ | $\cdots$ | $\cdots$ |
| | $o_{r1}$ | $o_{r2}$ | $\cdots$ | $o_{rc}$ | $n_r.$ |
| $c$ Total | $n_{.1}$ | $n_{.2}$ | $\cdots$ | $n_{.c}$ | $n$ |

Note that $n$ is the overall total, $n = \sum_{i=1}^{r} n_i. = \sum_{j=1}^{c} n_{.j}$

The $p$-value for testing $H_0$ that the row and column are independent (no association) against $H_1$ that there is association is $\Pr\{\chi^2_{(c-1)(r-1)} > Q\}$, where

$$Q = \sum_{i=1}^{r}\sum_{j=1}^{c} \frac{(o_{ij} - e_{ij})^2}{e_{ij}}; \quad e_{ij} = \frac{n_i. n_{.j}}{n}. \tag{A.4.1}$$

(iii) A special case of the $r \times c$ table is the $2 \times 2$ contingency table

| | Column 1 | Column 2 | Total |
|---|---|---|---|
| Row 1 | $o_{11}$ | $o_{12}$ | $n_1.$ |
| Row 2 | $o_{21}$ | $o_{22}$ | $n_2.$ |
| Total | $n_{.1}$ | $n_{.2}$ | $n$ |

From this table, the chi-square test for independence between the row and column is

$$Q = \frac{n(o_{11}o_{22} - o_{12}o_{21})^2}{n_1. n_2. n_{.1} n_{.2}} \sim \chi^2_1. \tag{A.4.2}$$

It can be shown that the test by (A.4.2) is equivalent to the two sample Z-test in comparison of two proportions, i.e, $Q = Z^2$, where

$$Z = \frac{|o_{11}/n_{.1} - o_{12}/n_{.2}|}{\sqrt{\hat{p}(1 - \hat{p})(1/n_{.1} + 1/n_{.2})}}, \tag{A.4.3}$$

where $\hat{p} = n_1./n$ is the pooled estimate of P(Row 1) under the null hypothesis. This equivalence is useful because Eq. (A.4.3) can be used to compute the power when the alternative is specified.

Since the chi-square test is asymptotic, one may use Fisher's exact test, conditional on the marginal totals, $n_{.1}, n_{.2}, n_1., n_2.$. The details can be found in Agresti (1990, p. 60).

# A.5   Discriminant Analysis and Neural Network

## A.5.1   Linear Discrimination Function based on Multivariate Normal Distribution

**Given**: Let the $x$s in this section be the same as those defined by (A.1.1). When data in Table A.5.1 are given as the learning sample, we wish to classify a new observation $x$ of unknown origin to one of the classes satisfying certain optimal criterion.

Table A.5.1: Classes, data, and prior probabilities for discriminant analysis.

| Class | Notation | Data | Prior |
|-------|----------|------|-------|
| 1 | $\omega_1$ | $x_{11}, x_{12}, \ldots, x_{1n_1}$ | $\pi_1$ |
| 2 | $\omega_2$ | $x_{21}, x_{22}, \ldots, x_{2n_2}$ | $\pi_2$ |
| $\vdots$ | $\vdots$ | $\vdots$ | $\vdots$ |
| $k$ | $\omega_k$ | $x_{k1}, x_{k2}, \ldots, x_{kn_k}$ | $\pi_k$ |

**To find**: The best way to classify $x$, and the error of misclassification.

**Basic theory**: Let $p(x|\omega_j)$ denote the density (if $x$ is continuous) or the probability (if $x$ is discrete) of $x$ when $x \in \omega_j$.

If $p(x|\omega_j)$ is known for $j = 1, 2, \ldots, k$, then the classification that minimizes the misclassification error is accomplished by assigning $x$ to the largest posterior distribution

$$f(\omega_j|x) = \pi_j p(x|\omega_j) \left/ \sum_{j=1}^{k} \pi_j p(x|\omega_j) \right. \tag{A.5.1}$$

$$\propto \pi_j p(x|\omega_j). \tag{A.5.2}$$

The function in the denominator of (A.5.1)

$$p(x) \equiv \sum_{j=1}^{k} \pi_j p(x|\omega_j) \tag{A.5.3}$$

is called the distribution function for all of the classes combined.

When the severities of misclassification are not equal, e.g., to assign a patient to the healthy group may not be as bad as to assign a healthy person to the patient group, we define a general loss function

$$\lambda(\omega_i|\omega_j) = \text{loss of classifying an element } x \in \omega_j \text{ to } \omega_i. \tag{A.5.4}$$

A special case of this, equal misclassification loss, is

$$\lambda(\omega_i|\omega_j) = \begin{cases} 1 & \text{if} \quad \omega_i \neq \omega_j \\ 0 & \text{if} \quad \omega_i = \omega_j. \end{cases} \tag{A.5.5}$$

Obviously, if we assign $x$ to $\hat{\omega}(x) = \omega_i$, then our loss is

$$\ell^i(x) = \sum_{j=1}^{k} \lambda(\omega_i|\omega_j) f(\omega_j|x).$$

Thus, the decision rule that minimizes the risk is to let

$$\hat{\omega}(x) = \omega_i \quad \text{if} \quad \ell^i(x) \leq \ell^j(x) \quad \text{for all} \quad j \neq i. \tag{A.5.6}$$

Note that this rule reduces to (A.5.2) if the loss function is (A.5.5). For any given $x$, the loss due to misclassification is

$$\ell^*(x) = \min_i \ell^i(x),$$

and the overall loss of this procedure is

$$L^* = \int \ell^*(x)p(x)dx. \tag{A.5.7}$$

In particular, if the loss function is (A.5.5), $L^*$ in (A.5.7) is the error of misclassification. □

## Some details

The prior distribution $\pi_j$'s are usually obtained in one of the following three ways:

1. Assuming $\pi_1 = \pi_2 = \cdots = \pi_k$.
2. If the data is a random sample from the population, then

$$\pi_j \simeq \frac{n_j}{\Sigma n_j}.$$

3. Estimating $\pi_j$ from other sources.

Sometimes it is more realistic to add one more classification class, called "no decision" $\omega_0$, i.e., if no decision is made on how to classify $x$, $x$ is assigned to $\omega_0$. In this case, one has to specify all the losses

$$\lambda(\omega_0|\omega_j), \quad j = 1, 2, \ldots, k,$$

and modify the rule of (A.5.6), e.g. if the $\min_i(\ell^i(x)) > \ell_0$ for a given $\ell_0$, assign $x$ to $\omega_0$.

When $p(x|\omega_j)$ is assumed to have a multivariate normal density with a common covariance matrix, i.e. $\omega_j \sim N(\mu_j, \Sigma)$, the data can be used to estimate $\mu_j$ and $\Sigma$. The optimal classification rule under loss function (A.5.5) is assign $x$ to $i$ if

$$(x - (\hat{\mu}_i + \hat{\mu}_j)/2)'\hat{\Sigma}^{-1}(\hat{\mu}_i - \hat{\mu}_j) > \ln(\pi_j/\pi_i), \tag{A.5.8}$$

for all $j$, where

$$\hat{\mu}_i = \sum_{j=1}^{n_j} x_{ij}/n_i, \quad \hat{\Sigma} = \frac{\sum_{i=1}^{k}\sum_{j=1}^{n_i}(x_{ij} - \bar{x}_i)(x_{ij} - \bar{x}_i)'}{\sum_{i=1}^{k} n_i}.$$

## A.5.2   Neural Network for Discrimination

A neural network tries to mimic the function of neural cells and their connections. Fig. A.5.1(A) represents the functions of neurons in a living organism. The thin lines are the dendrites which receive signals for the neural cells (circles). The neural cells process the incoming information and make a decision whether to send a signal to the next level through the exons (thick arrows). Fig. A.5.1(B) is a typical neural network with one "hidden" layer. More complicated neural networks may have additional hidden layers. The similarity between Fig. A.5.1(A) and (B) is easy to see. The input $x$'s are first assembled in the first layer neurons with the output $\boldsymbol{\psi}(\boldsymbol{\alpha}'\chi)$ where $\boldsymbol{\psi}(\cdot)$ is a vector of functions and $\boldsymbol{\alpha}$ are weights for the $x$. The next layer summarizes the input $\boldsymbol{\psi}$ and output

$$y = \phi(\boldsymbol{\beta}'\boldsymbol{\psi}), \qquad (A.5.9)$$

where $\phi$ is another function and $\boldsymbol{\beta}$ are other weights. The functions $\boldsymbol{\psi}$, $\phi$, the number of layers and the number of neurons in each layer are determined by the network designer, but the weights can be tuned during learning. There seems to be no strict rule for an optimal design. Most times, the best design has to be determined by trial and error. Moreover, each neuron does not have to receive all the signals from the previous layer. Some of them can specialize in certain input subsets. The functions

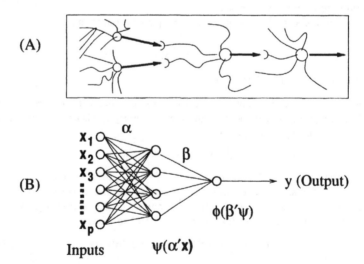

Figure A.5.1:  (A) Neurons and their connections. The thin lines denote neuron dendrites where information is gathered, the circles denote neuron cell bodies where information is processed, and the thick arrows denoted the axons where output from the cells is sent. (B) A neuron network. The inputs are first processed by $\boldsymbol{\psi}(\boldsymbol{\alpha}'X)$ at the nodes in the "hidden" layer and then at the output neuron using $\phi(\boldsymbol{\beta}'\boldsymbol{\psi})$. There can be more than one hidden layer between the input and output.

$\psi$ and $\phi$ can have many choices. Since a neuron has output of either 0 or 1, a natural function to mimic a neuron is to let $\psi_i(\boldsymbol{\alpha}_i' \boldsymbol{X}) = 0$ if $\boldsymbol{\alpha}_i' \boldsymbol{X} < \zeta_i$, and 1 otherwise, where $\zeta_i$ is the threshold for the $i$th neuron, which has a weighting vector $\boldsymbol{\alpha}_i$. Since step functions are not easy to work with, a commonly used function is the sigmoid function.

$$\psi(a) = \frac{1}{1 + e^{-a}}. \tag{A.5.10}$$

For large $\boldsymbol{\alpha}$ weights, one can almost get a 0, 1 step function.

A neural network "learns" by adjusting the weights from a training sample. To simplify notation, we may rewrite (A.5.9) as

$$y = f(\boldsymbol{x}, \boldsymbol{\theta}), \tag{A.5.11}$$

where $\boldsymbol{\theta}$ contains all the weights of $\boldsymbol{\alpha}$s and $\boldsymbol{\beta}$s. Suppose the $n$ training sample has been used and the prediction function for $y$ is

$$y = f(\boldsymbol{x}, \boldsymbol{\theta}_n). \tag{A.5.12}$$

Suppose datum $n + 1$ with input $\boldsymbol{x}_{n+1}$ and output $y_{n+1}$ is the next learning sample, when $\boldsymbol{x}_{n+1}$ is put into the current prediction equation (A.5.12), the predicted value is

$$\hat{y}_{n+1} = f(\boldsymbol{x}_{n+1}, \boldsymbol{\theta}_n).$$

Clearly, if $\hat{y}_{n+1} = y_{n+1}$, $\boldsymbol{\theta}_n$ should be kept, i.e. $\boldsymbol{\theta}_{n+1} = \boldsymbol{\theta}_n$. Otherwise $\boldsymbol{\theta}_n$ should be modified according to the difference $y_{n+1} - \hat{y}_{n+1}$. Most neural networks (White, 1989, Fu, 1994) use the formula

$$\boldsymbol{\theta}_{n+1} = \boldsymbol{\theta}_n + \eta \frac{\partial f}{\partial \boldsymbol{\theta}_n} (y_{n+1} - \hat{y}_{n+1}), \tag{A.5.13}$$

where $\partial f / \partial \boldsymbol{\theta}_n$ is the vector containing all the partial derivative of $y$ with respect to each member in $\boldsymbol{\theta}$ evaluated at $\boldsymbol{\theta}_n$, and $\eta$ is a pre-assigned positive number. The reason for using (A.5.13) can be seen from Taylor's expansion of $f(\boldsymbol{x}, \theta + \Delta\theta)$ for univariate $\theta$. Because we wish $f(\boldsymbol{x}_{n+1}, \theta + \Delta\theta) = y_{n+1}$, we need

$$f(\boldsymbol{x}_{n+1}, \theta) + f'(\boldsymbol{x}_{n+1}, \theta)\Delta\theta \approx y_{n+1}, \quad \text{or}$$

$$f'(\boldsymbol{x}_{n+1}, \theta)\Delta\theta \approx (y_{n+1} - \hat{y}_{n+1}),$$

where $f'(\boldsymbol{x}_{n+1}, \theta)$ is the partial derivative with respect to $\theta$. Apparently $\Delta\theta = \theta_{n+1} - \theta_n$ has the same sign as $f'(\boldsymbol{x}_{n+1}, \theta)(y_{n+1} - \hat{y}_{n+1})$. Instead of using the increment $(y_{n+1} - \hat{y}_{n+1})/f'(\boldsymbol{x}_{n+1}, \theta)$ for $\Delta\theta$ in Newton's method, only a small increment $\eta f'(\boldsymbol{x}_{n+1}, \theta)(y_{n+1} - \hat{y}_{n+1})$ is used to avoid drastic changes when $f'(\boldsymbol{x}_{n+1}, \theta)$ is small. It has been shown by White (1989) that under certain regularity conditions, $\boldsymbol{\theta}_n$ will either converge to $\boldsymbol{\theta}^*$ with probability 1 or to $\infty$ with probability 1, where $\boldsymbol{\theta}^*$ satisfies $E[\partial(y - f(\boldsymbol{x}, \boldsymbol{\theta}))^2 / \partial\boldsymbol{\theta}] = 0$ with the $E[\,]$ representing the expectation with respect to the $(\boldsymbol{x}, y)$ distribution. In other words, a neural network may either diverge or converge to a local minimum in prediction mean square error.

# Appendix B

# Selected Public Domain Computer Programs for Genetic Data Analysis

A list of computer programs compiled by Wentian Li at the Laboratory of Statistical Genetics at Rockefeller University, New York, contains many public domain genetic data analysis computer software. The following computer programs, except those by the author of this book, can be found in his list (referred as the Li's list). To access the list use

$$http://linkage.rockefeller.edu/soft/list.html$$

When a public domain program is discussed here, we explain only how to fetch it, how to compile it and how to get started with a simple example. To get to all the the features of a program requires reading the manual. A good starting point is to get at least one data set correctly analyzed. This is what this appendix tries to do. Simple examples are provided in Yang's web site

$$http://www.stat.ufl.edu/\sim yang/genetics$$

## B.1    A Teaching Linkage Analysis Program for Two Generation Families

A computer program has been written for two generation families like the two in Fig. B.1.1. This program is used to understand the construction of the likelihood function for linkage analysis.

Table B.1.1: Possible positions $(1, 2, \ldots, 16)$ of an offspring
from a general A1/a2 × A3/a4 phase-known mating.

|  | | Father: A1/a2 Phase-known | | | |
|---|---|---|---|---|---|
|  | | A1 | a2 | A2 | a1 |
|  | A3 | 1 | 2 | 3 | 4 |
| *Mother:* | a4 | 5 | 6 | 7 | 8 |
| A3/a4 | A4 | 9 | 10 | 11 | 12 |
| *Phase-known* | a3 | 13 | 14 | 15 | 16 |

Only one marker locus can be considered in this program. The number of alleles of the marker can be any number and the phase relation at the parental level can be known or unknown. This program can calculate the LOD score of a single family with two parents and any number of children. It may be used for a large pedigree if the pedigree can be dissected into many two generation ones. The locus that may cause a disease is assumed to have two alleles, $A$ and $a$, and the disease allele can be dominant or recessive. The recombinational fractions can be the same or different in the paternal and maternal sides. What a user needs to do is to determine the **inseparable class** for each offspring and input them into the computer. Let us use Table B.1.1 for illustration.

This table covers all of the cases of parents' genotype in finding the linkage between the disease gene and the marker. An inseparable class is defined as the entries in the table that are inseparable given the offspring's phenotype and marker information. For example, suppose the marker information for the two parents is the four different alleles given in the table and the disease allele $A$ is dominant. Suppose a son has marker 14 and he is affected. Then he can be in positions 5, 9 or 12 in the table. There is no way to tell which one he is. Thus the three status 5, 9, 12 form an inseparable class for him. On the other hand, if he is an unaffected person with marker 14, then he has to be case 8, or his inseparable class has only one element 8. The need to identify these inseparable classes is obvious because they jointly define the probability that this offspring can be produced (see Table 2.2.1).

Now suppose the father is still $A1/a2$, but the mother is $a2/a2$ in a double backcross mating. We still assume that $A$ is the dominant disease gene. To find the inseparable classes for a child, the user needs to replace all of the 4 maternal haplotypes by $a2$. Consequently, an unaffected child with marker 22 must be 2, 6, 10, 14 and an affected child with marker 22 must be 3, 7, 11, 15. These numbers determine his/her inseparable class in Table B.1.1. When the parents' phase in unknown, the user needs only to do one of the two possible phases and tell the computer that the phase is unknown. The program will take care of the other phase possibility.

This program is available as *link2.f, link2.exe, link2.dat* and *link2.out* at Yang's web site *AppB*.1.

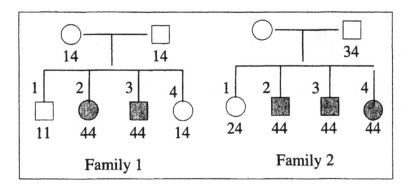

Figure B.1.1: Example of two families for linkages analysis for Example B.1.1.

**Example B.1.1** *Determine the inseparable classes for the offsprings in the two families in Fig. B.1.1. Note that these are the same as the pedigrees in Fig. 2.1.4.*

**Sol**: For Family 1, Table B.1.1 can be written as

|  |  | Father: A1/a4 | | | |
|---|---|---|---|---|---|
|  |  | A1 | a4 | A4 | a1 |
|  | A1 | 1 | 2 | 3 | 4 |
| Mother: | a4 | 5 | 6 | 7 | 8 |
| A1/a4 | A4 | 9 | 10 | 11 | 12 |
|  | a1 | 13 | 14 | 15 | 16 |

Apparently, the disease gene is recessive. Thus, the inseparable class for offspring 1 is 1, 4, 13, and for 2 and 3 are 6 and for 4 is 2, 3, 5, 9, 12 and 15. They are in the inputs in *link2.dat*. The inseparable classes for Family 2 are also given in *link2.dat*.    □

The *link2.dat* file (for input to *link2.exe*):

```
---------------------------------------------------
 2 6           (2I2)          Saved as link2.dat
0.000 0.000    (2F6.3)
0.050  0.050
0.100 0.100
0.150 0.150
0.200 0.200
```

```
0.500 0.500
 4 0 0          (3I2)
1.0 0.0 0.0 1.0 0.0 0.0 0.0 0.0 0.0 0.0 0.0 0.0 1.0 0.0 0.0 0.0  (16F4.1)
                1.0
                1.0
0.0 1.0 1.0 0.0 1.0 0.0 0.0 0.0 1.0 0.0 0.0 1.0 0.0 0.0 1.0 0.0
 4 0 0
0.0 0.0 0.0 0.0 1.0 0.0 0.0 0.0 1.0 0.0 0.0 1.0 0.0 0.0 0.0 0.0
                1.0
                1.0
                1.0
```

Input instructions:


Line 1:    # of families=2, # of theta to be evaluated
    2-7:    theta of father's side, theta of mother's side
    8:      Size of Family 1, father's phase*, mother's phase*
    9-12:   Inseparable class for each child**
    13:     Size of Family 2, father's phase, mother's phase
    14-17:  Inseparable class for each child


    *When phase is known,  use 1, otherwise use 0.  For example,
     1 0 means father's phase is known, but mother's phase is unknown.


    **The 1s denote the inseparable class for this individual. The
      16 positions are for the 16 positions in Table B.1.1.
- - - - - - - - - - - - - - - - - - - - - - - - - - - - - - - - - - - - - - - - - -


The outputs from this program *link2.out* are listed below. Results are separated
for the individual family and a summary.
- - - - - - - - - - - - - - - - - - - - - - - - - - - - - - - - - - - - - - - - - -

FAMILY    1


| R MALE | R FEMALE | LOD-SCORE |
|--------|----------|-----------|
| 0.00000 | 0.00000 | 0.85194 |
| 0.05000 | 0.05000 | 0.74193 |
| 0.10000 | 0.10000 | 0.62994 |
| 0.15000 | 0.15000 | 0.51764 |
| 0.20000 | 0.20000 | 0.40724 |
| 0.50000 | 0.50000 | 0.00000 |

```
FAMILY    2
```

|  R MALE  |  R FEMALE  |  LOD-SCORE  |
| --- | --- | --- |
| 0.00000 | 0.00000 | 1.32906 |
| 0.05000 | 0.05000 | 1.17434 |
| 0.10000 | 0.10000 | 1.01433 |
| 0.15000 | 0.15000 | 0.84977 |
| 0.20000 | 0.20000 | 0.68252 |
| 0.50000 | 0.50000 | 0.00000 |

```
OVERALL Z VALUES
```

|  R MALE  |  R FEMALE  |  LOD-SCORE  |
| --- | --- | --- |
| 0.00000 | 0.00000 | 2.18100 |
| 0.05000 | 0.05000 | 1.91627 |
| 0.10000 | 0.10000 | 1.64427 |
| 0.15000 | 0.15000 | 1.36741 |
| 0.20000 | 0.20000 | 1.08975 |
| 0.50000 | 0.50000 | 0.00000 |

```
The three values in each row are theta on father's side, theta on mother
side, LOD score for the linkage between the disease gene and the marker.
```

# B.2  A Simulation Program to Compute ELOD and Var(LOD) for Two Generation Families

This program can calculate the expected LOD score and its variance for a single family with two parents and siblings when the mating types, such as double backcross, are known. The concept of inseparable class in Table B.1.1 is used. This program input and output examples are available as *psim.f*, *psim.dat* and *psim.out* at Yang's web site *AppB.2*. The input instruction is given in the comments of the program *psim.f*. The data *psim.dat* file is for the double backcross experiment in Table 2.4.1. The output includes the result (2.4.4).

A. Input *Tab241.dat*

- - - - - - - - - - - - - - - - - - - - - - - - - - - - - - - - - - - - - - - - - - - - - -

```
4   2   5000   12345.0      (2I3,I6,f10.1)   Saved as psim.dat
4   1   5  9 13             (17I3)
4   2   6 10 14
4   3   7 11 15
4   4   8 12 16
5                           (I3)
.00   .00 0   .00   .00 0   (2F5.1, I2, 2F5.1, I2)
0.1   0.1 0   0.1   0.1 0
.15   .15 0   .15   .15 0
0.2   0.2 0   0.2   0.2 0
.20   .10 0   .20   .10 0
```

Input instructions:

Line 1:    # of inseparable classes, # of children, simulation size, seed
           for simulation (integer in real number format)
    2-5:   Size of the inseparable class and their numbers in Table B.1.1*
           One line for each inseparable class
           Line 2: inseparable class for Aa/11
           Line 3: inseparable class for aa/12
           Line 4: inseparable class for Aa/12
           Line 5: inseparable class for aa/11
    6:     Number of theta to be evaluated
    7-11:  For each line, there are six input numbers.
Father's true theta, father's theta to be evaluated, his phase status,
Mother's true theta, mother's theta to be evaluated, her phase status.
                (For phase status, 0 for unknown, 1 for known.)

*       Table B.1.1 for this case Aa/12 x aa/11

|     | A1 | a2 | A2 | a1 |
|-----|----|----|----|----|
| a1  | 1  | 2  | 3  | 4  |
| a1  | 5  | 6  | 7  | 8  |
| a1  | 9  | 10 | 11 | 12 |
| a1  | 13 | 14 | 15 | 15 |

- - - - - - - - - - - - - - - - - - - - - - - - - - - - - - - - - - - - - - - - - - - - - -

A. Output file *psim.out* (partial): Use the second line as an example: If the father's (male's) true recombination fraction $\theta = 0.1$ and the $\theta$ is to be evaluated at 0.1 and his phase information is unknown and the same condition holds for the mother's side, then this pedigree will give a mean LOD 0.09552 with standard deviation 0.25369 (or variance 0.06436).

File *Tab241.out*

---------------------------------------------------------------

| TF | TFE | IPF | TM | TME | IPM | MEAN | VAR | SD |
|------|------|-----|------|------|-----|---------|---------|---------|
| 0.00 | 0.00 | 0 | 0.00 | 0.00 | 0 | 0.30103 | 0.00000 | 0.00000 |
| 0.10 | 0.10 | 0 | 0.10 | 0.10 | 0 | 0.09552 | 0.06436 | 0.25369 |
| 0.15 | 0.15 | 0 | 0.15 | 0.15 | 0 | 0.05324 | 0.04147 | 0.20364 |
| 0.20 | 0.20 | 0 | 0.20 | 0.20 | 0 | 0.03036 | 0.02313 | 0.15210 |
| 0.20 | 0.10 | 0 | 0.20 | 0.10 | 0 | 0.00701 | 0.09369 | 0.30608 |

TF,TFE,IPF = Father's true theta, evaluated theta, phase information
TM,TME,IMF = mother's true theta, evaluated theta, phase information
MEAN, VAR, SD = Mean, variance, standard deviation for LOD for this
                type of family

---------------------------------------------------------------

# B.3 *Liped* and Programs Described in Terwilliger and Ott's Handbook

Terwilliger and Ott's *Handbook of Human Genetic Linkage* (1994) has listed many linkage programs and they can be obtained from the authors free of charge. The programs include:

1. Linkage program for general pedigree with dominant, recessive and sex-linked diseases.
2. Disease with age-dependent penetrance.
3. Multipoint linkage analysis (Multi-loci markers). In this general area, markers with unlikely linkage to the gene (e.g., the region $\{\theta : Z_\theta < -2\}$) are suggested and should be excluded for future typing to save resources.
4. When the recombination rates show sex difference.
5. When interference occurs between loci (see Exercise 1.8).
6. Mutation rate is considered in linkage analysis.
7. Gene mapping based on linkage disequilibrium.
8. Affected sib-pair analysis.
9. Pedigree simulation.

The linkage program can be fetched from *ftp* : *//linkage.rockefeller.edu/software /linkage*. However, the input to this package involves several steps and its explanation is beyond the scope of this book. Here we consider only **liped**. It is relatively easy to use and can analyze reasonable size pedigrees. The steps to get started are:

1. Ftp the file from *http* : *//linkage.rockefeller.edu/software/liped* (Manual can be found from *http* : *//linkage.rockefeller.edu/ott/liped.html*.)

2. Find the version that fits your system (UNIX is used as example).
3. Uncompress the file (called *liped.tar.Z*)
4. Now you see files: *liped.for constant.inc, subprog.inc, examples.dat*
5. Compile the fortran program (In Unix, rename *liped.for* as *liped.f*, use f77 compiler. Now you see the executable file *a.out*. Change its name to *liped.exe*).
6. Use the data set in Yang's web site *AppB*.3. Change the file *FigB*11.*dat* into *liped.dat* (*liped* used the file with name *liped.dat* (a description is given after each line)). Run *liped.exe*, you will see *liped.out* which is the output file. The results should agree with the output from *link2.out* from the previous section. Try again with *Fig*113.*dat*. The output should be agree with what described in Section 1.1. Special attention should be paid to a pedigrees like Fig 1.1.3, which is defined a complex pedigree and the method of data input has to be modified. Description is given in the *Fig*113.*dat*.
7. More examples are provided in *examples.dat* from the *liped* package. *Liped* can handle sex-linked inheritance, multi-marker loci, mutation and age dependent penetrance.
8. *Liped* can also handle quantitative trait by (2.2.17) in § 2.2.1.

# B.4   Simlink—Power Prediction for Given Pedigrees

Simlink, written by Ploughman and Boehnke, has been available for the genetics community since 1990. It estimates the power for untyped pedigrees when the potential families for the gene hunting are known. Here is how to get started with this program.

1. Simlink can be fetched from the instruction at *http* : //*www.sph.umich.edu/ group/statgen/software*. Manual can be found from online documentation at *SIMLINK* from *http* : //*linkage.rockefeller.edu/soft/list.html*.
2. Find the version that fits your system. UNIX is used as example.
3. Uncompress the file (called *simlink.tar*).
4. Now you see files: *simlink, simlinki.f*, etc.
5. The file simlink is an executable file. If it does not work for your system, you need to compile *simlink.f sim1.f sim2.f men1.f men2.f* together to create an executable file.
6. Use the data set *Fig*243.* in Yang's web site *AppB*.4. to run the executable file. Or you may start with the examples in the *Simlink* package. Input description is given in the input files. Of course, reading the manual is necessary. The files *Fig*243.* are for Fig. 2.4.3 when we discuss the Phoughman and Boehnke (1989) paper.

7. The program is interactive. When the executable file is run, it asks only two files, the control file, which is *Fig243.con*, and the output file, which is the name you type in.

8. Caution: The *.ped* and *.loc* files have a hidden end-of-file symbol at the end. Your wordprocessor may not have this symbol. It will then cause error. The best way is to copy the example files and modify them.

9. The outputs include expected mean LOD score and standard deviation at various values of $\theta$s (in Table 2 in *Fig243.out*), the probability that the LOD will be greater than a threshold (in Table 3 in *Fig243.out*), and expected LOD under $H_0$ (in Table 4 in *Fig243.out*) and the probability that the LOD will be greater than a threshold under $H_0$ (in Table 4 in *Fig243.out*).

10. Example 2 in the package represent Fig 2.4.4.

11. *Simlink* can also handle quantitative trait.

# B.5 SIB—PAIR: A Nonparametric Linkage Software

SIB—PAIR, written by David Duffy, contains many affected sibpair and affected pedigree methods. This package provides many important pedigrees in the literature. Here is how to get started.

1. Get the right file for your computer system from *http* : //*www.qimr.edu.au /davidD/davidd.html* (in Li's list). For example, *sib-pair.f.gz*. Uncompress *sib-pair.f.gz* and then compile this fortran program. Now you have the executable program, say, *sibpair.exe*.

2. Also fetch the example file *examples.tar.gz* and uncompress it. It provided many pedigree examples (in *PEDIGREE*) and input instructions (in *SCRIPT*) for *sibpair.exe*. The manual is the *sib-pair.ps* file.

3. Use the file *Fig234.** in Yang's web site *AppB*.5 as an example.

4. Run the program as *sibpair.exe < Fig234.in > Fig234.out*, i.e., to run the input file *Fig234.in* and output it in *Fig234.out*. This input represents the pedigree in Fig 2.3.4.

5. There are many outputs. We use the Whittemore and Halpern test discussed in §2.3.4 as an example. The score function used in this software is $\Phi_2$. At the "Identity-by-descent based statistic" you will see a row of Whit-Halp Score $E(Z) = 0.555$, $Var(Z) = 0.072$ and $Z = 0.50$, which correspond to $E[T_1|F_1^c, H_0]$, $V[T_1|F_1^c, H_0]$ and $T_1$ respectively in Table 2.3.11. Note that the mean and variance in *Fig234.out* were obtained by simulation. Hence, they are not exactly the same as the theoretical values given in Table 2.3.11. Actually, the values vary each time the program is run.

6. The *Exp*311.∗ files are for the Haseman-Elston test. The data are given in Example 3.1.1. This output is in the H-E for trait "trait" v. locus "marker" section. Note that the equation used in this software is not the same as (3.1.13). It is $\Delta^2 = \alpha + \beta(\hat{\pi}_m/2) + \varepsilon$. The estimated slope doubles, but the p-value based on the t-test is the same.

# B.6   MAPMAKER/SIBS, MAPMAKER/QTL, Genehunter, GAS (Genetic Analysis System)

The four programs listed in the title of this section use similar input output format. All of them deal with multiple loci sliding window type genehunting with numerical and graphics outputs. All of them can be fetched from Li's list. Only *MAPMAKER/SIBS*, written by Kruglyak, Daly and Lander, will be discussed here. To get started

1. The program can be obtained from

   *ftp − genome.wi.mit.edu/distribution/software/sibs*

   After uncompress *sib-2.1.tar.Z*, you see *sibs.sun*, *Makefile*, etc.
2. Change *Makefile* according to your system. For example, for SUN Solaris, change SYS = -D_SYS_SUNOS and CC = cc to

   SYS = -D_SYS_SOLARIS

   CC = gcc

   and save the file. Then run *Makefile*, you see the executable file *sibs*.
3. When you run *sibs*, you'll see prompt

   sibspair:1>

   asking for an instruction.

   The first instruction should be to *load markers* from a loc file, e.g.,

   sibspair:1>load markers E321.loc

   You will see: sibspair:2>.

   You may put in the pedigree information

   sibspair:2>prepare pedigrees E321.ped

   You will see: Load phenotype data? y/n [n]:

   If you put *y*, you need to provide the file for phenotype information, e.g. *EX*321.*pheno*. The next instruction is:

   sibspair:3>scan

   You will see: Scan pairs ...? p/a[p]: (p for phenotype data, a for affectation data)

   Now there are many options for QTL data: *haseman elston, nonparametric, ml variance* and options for affected status data: *estimate, informap*.

The files under $E321$ in Yang's web site *AppB*.6, are the same data in Example 3.2.1. The LOD scores from this program for Haseman-Elston and mle variance test are the same as the values in the solution of the example. The Z-scores for the non-parametric have some discrepancy (1.940 versus 1.988). The reason may be due to the variance used to compute the $Z$ score is exact in the example but is only an approximation in the software (see the 4th paragraph of page 450 of Kruglyak and Lander, 1995). The map function used in MAPMAKER software is the Kosambi's function (Ott, 1991), i.e., the physcial distance $x$ and the recombination fraction $\theta$ are related by

$$x = \frac{1}{4}\ln\left(\frac{1 + 2\theta}{1 - 2\theta}\right).$$

For example, the input recombination fractions 0.4 and 0.2 in $E321.loc$ are converted to 54.9 cM and 21.18 cM in the output graphics.

# B.7 A SAS Program to Compute Kinship Coefficients

This SAS program applied to Example 3.3.4.

```
/* ###### Computer Program ###### */
data EX334;
infile 'ex334.ped';
input animal sire dam;

proc inbreed data=EX334 matrix covar outcov=amat;
title1 '*** proc inbreed output ***';
var animal sire dam;
data _null_;set amat;
file 'ex334.out';
put @1 col1 6.4 @8 col2 6.4 @15 col3 6.4 @22 col4 6.4
@29 col5 6.4 @36 col6 6.4 @43 col7 6.4;
run;

/* ###### Input data at ex334.ped ###### */
/* data are in the order individual father mother */

01 . .
02 . .
03 01 02
04 01 02
```

```
05 .  .
06 03 04
07 04 05
```

```
/*  ###### Output at ex334.out #####  */
```

```
1.0000 0.0000 0.5000 0.5000 0.0000 0.5000 0.2500
0.0000 1.0000 0.5000 0.5000 0.0000 0.5000 0.2500
0.5000 0.5000 1.0000 0.5000 0.0000 0.7500 0.2500
0.5000 0.5000 0.5000 1.0000 0.0000 0.7500 0.5000
0.0000 0.0000 0.0000 0.0000 1.0000 0.0000 0.5000
0.5000 0.5000 0.7500 0.7500 0.0000 1.2500 0.3750
0.2500 0.2500 0.2500 0.5000 0.5000 0.3750 1.0000
```

This program is available as *Ex334.sas*, *Ex334.ped* and *Ex334.lst* at *AppB*.7 Yang's web site.

# B.8  Using SAS-IML to Solve Henderson's Equation

The program *Ex341.sas* and inputs *Ex341.ped* and *Ex341.dat* saved in Yang's web site *AppB*.8 is to solve Example 3.4.1, but it can be modified to similar programs, provided the reader knows SAS-IML. The output is saved as *Ex341.lst*.

The following discussion is based on the listed comments in program *Ex341.lst*. The page numbers refer to the same output.

A. Enter the pedigree (*Ex341.ped*) of the animals to let SAS PROC INBREED compute the kinship matrix. Also, enter the trait values from the table given in the example (*Ex341.dat*).

B. In a case where the kinship matrix is known and the $\sigma_a^2/\sigma_e^2$ ratio is known, this is how to input the data. The ratio $\sigma_a^2/\sigma_e^2 = 5.43$ is used in this example. Under this condition, PROC MIXED can be directly used and the output is in the first two pages of the SAS output *Ex341.lst*. The output from PROC MIXED program under this condition B is on pp. 1–2.

C. When using a pedigree to find kinship matrix, the output is on pp. 3–6.

D. Form the design regression matrix for IML program.

E. Maximize the likelihood equation (3.4.6). Output on p. 7. Note that $\sigma_s^2 = 296$ and $\sigma_e^2 = 65$ maximizes the likelihood function. The log likelihood at these two optimal points is $-46.37241$.

F. The estimated $\sigma_s^2 = 296$ and $\sigma_e^2 = 65$ are put into PROC MIXED. The outputs are on pp. 8–9. The sire and herd effects are given in the outputs.

# B.9 ETDT (Extended Transmission Disequilibrium Test)

ETDT, written by D. Curtis, can do several transmission disequilibrium tests discussed in § 4.1. An executable PC(DOS) program package can be fetched from web site:

$$ftp : //ftp.gene.ucl.ac.uk/pub/packages/dcurtis.$$

The manual is *programs.txt* in the package.

The original batch program *etdt.bat* starts with a pedigree preparation program *ped2tdt.exe* by assuming that the pedigree is written in the form of *example.ppd*. It is easier if one starts with a pedigree like the one in *example.tdt*.

The *test.\** files given in Yang's web site *AppB*.9, start with a simple sib pedigree file *test.tdt*. In this file, the first line is the number of alleles followed by one line per affected child. The format is: the first two numbers are father's two alleles, the next two numbers are mother's two alleles, and then followed by the two alleles by the affected child. For example the second line

$$0\ 0\ 1\ 2\ 1\ 3$$

in *test.tdt* means that the father's alleles are unknown, the mother's are *1 2* and the affected child's are *1 3*; and the second line

$$1\ 1\ 1\ 3\ 1\ 3$$

means the father's alleles are *1 1*, mother's are *1 3* and the affected child's are *1 3*. The comments followed data show how etdt handle ambiguous and erroneous data. The outputs are *test.chi*, which gives the three TDT test 2\*(L1-L0), 2\*(L2-L0) and 2\*(L2-L1) discussed at the end of § 4.1 following (4.1.19), and *test.cou*, which gives the transmission history of the alleles.

Note that the batch file *etdt.bat* in the author's directory has been change, because the program *ped2tdt.exe* is not used. A comparison of this and the original *etdt.bat* reveals that the line before

calctdt %1.tdt %1.cou

has been removed. Also, it it apparent that when this program is run, the input data must be in the form of *.tdt*. For example to run *test.tdt*, use ">run etdt test".

# B.10 GeneBank and Genetic Data Sources

GeneBank has publicly accessible genetic data that can be retrieved in the following manner. Here we use the data cited in Buldyrev et al. (1993, Table 2) as an example.

The data are from bacterium *Bradyrhizobium japonicum* with length 703 bp and GenBank accession # M77796.

**Type** *http : //www.ncbi.nlm.nih.gov* and hit the return key. (*xmosaic* is needed before http:// if from terminal)

(You will see a window starting with **Welcome to NCBI**. Click on the words
• Searching GenBank.)

(Now you see the National Center for Biotechnology Information. Click on Entrez Browser)

(You see the **Entrez** window. Click at • nucleotides column) (Here, you see many options in Search Field. In our example, use accession. Click this column.)

(Now you type the gene you want, for example, M77796 in the pink window. After typing, hit the return key.)

(You may see a Security Warning box, press on Continue Submission.)

(You see the Current Query window. Click the [Retrieve 1 Document] window.)

(You may see another Security Warning box, press on Continue Submission.)

(Now you see the report option column in View GenBank report FASTA report ... [GenBank report gives you sources and biological information on the data, FASTA report: Data only is good for data entry.])

(If you have no preference, click on GenBank report)

(Click Display, you see the data.)

(Click [Save] box at the bottom of the screen.)

Your system should be able to save the genetic information file for M77796.

# Answers or Hints to Exercises

## Exercise 1

1.1  $L(\theta) \propto 2\binom{8}{4}\theta^4(1-\theta)^4 + \theta^8 + (1-\theta)^8$;  $L(0) > L(1/2)$.

1.2  (ii) (a) If $\sigma_1 > \sigma_2 > \sigma_3$, $p_n \to 1$; (b) If $\sigma_3 > \sigma_2 > \sigma_1, p_n \to 0$; (c) If $\sigma_2 > \sigma_1$ and $\sigma_2 > \sigma_3$, then $p_n$ oscillates between 0 and 1; (d) other conditions, $\sigma_1 > \sigma_3 > \sigma_2$, or $\sigma_3 > \sigma_1 > \sigma_2$, the convergence of $p_n$ depends on the initial condition.
(iii)  $p_n = p_{n-1}/(1 + p_{n-1}) = p_0/(1 + np_0)$. To make $p_n/p_0 = 0.5$, $n = 100$, or 3000 years.

1.3  (i) 1/2; (ii) 0.00025.

1.4  Similar to the albino example in §1.2.1

1.5  Since large proportions give more accurate estimate, we may let $\hat{p}_O = \sqrt{0.431} = 0.6565$, Solving $p_A^2 + 2p_A p_O = 0.422$, we have $\hat{p}_A = 0.267$. Use $\chi^2$ test

$$Q = \frac{0^2}{431} + \frac{0^2}{422} + \frac{(114 - 106)^2}{106} + \frac{(40 - 37)^2}{40}$$
$$= 0.828 \sim \chi_1^2.$$

There is no lack of fit.

1.6  (i) The chi-square values for 7 characteristics in Table 1.2.1 are respectively from 1 to 7; 0.262, 0.015, 0.391, 0.064, 0.451, 0.350, and 0.606, each with 1 d.f. Thus the overall test for the data being too good to be true has the $p$-value

$$p\text{-value} = Pr\{\chi_7^2 < 2.14\} \sim 0.05.$$

(ii) $Pr\{Aa$ is misclassified as $AA\} = (3/4)^{10} = 0.0563$. Hence the true ratio should be $(1 + 2 \times 0.0563) : 2(1 - 0.0563) = 1 : 1.696$.

1.7  (i) It is 0.38. (ii) can be proven for any two $x_1$ and $x_2$ with $\theta_i = (1 - e^{-2x_i})/2$, $i = 1, 2$.

1.9  Average length $= 4^6 = 4 \times 10^3$. $Pr\{$ fragment $> 10^4\} = \exp(-10^4/(4 \times 10^3)) = 0.082$. Total number $\simeq 3 \times 10^9/(4 \times 10^3) \times 0.082 = 6 \times 10^4$.

1.10  The length of chromosome $\simeq 300 \times 20\overset{\circ}{A} = 6 \times 10^{-5}$ cm. Thus a magnification of $10^4 \sim 10^5$ is necessary to see a chromosome in cm range.

1.11  Male color-blindness rate $= 725/9049 = 0.0801$. Thus, the female rate, according to sex-linked rule, should be $0.08^2$, or with 58.23 expected observations. By the Poisson distribution, 40 observed color-blind females was too low. (The reason turns out to be that there are two genes on $X$ chromosome that can suppress the expression of the color-blindness gene.)

## Exercise 2

**2.1**   300.

**2.2**   Chromosome 22 has length 56 the proportion of which is $56/3200 = 0.0175 \ll 1/22$. Thus, we should let

$$h(\theta) = \begin{cases} 1/2 & \text{with probability } 0.9825 \\ U(0, 0.5) & \text{with probability } 0.0175 \end{cases}$$

**2.3**   Let $f_1 = f_2 = m_1 = m_2 = (1 - \theta)/2$, $f_3 = f_4 = m_3 = m_4 = \theta/2$. Then

$$L(\theta) = (m_2 f_4)^2 m_4 f_4 [m_1 f_1 + m_1 f_4 + f_1 m_4]^2 [m_3 f_2 + m_2 f_3 + m_3 f_3]^2$$

**2.4**   Let $z$ be their physical distance, then the density function for $z$ is $f(z) = 2z/L^2, 0 < z < L$. $f(\theta) = -\ln(1 - 2\theta)/[L^2(1 - 2\theta)], 0 < \theta < (1 - e^{-2L})/2$.

**2.5**   For family 1, $L_1 \propto (1 - \theta)^5(\theta^3 + 1) + \theta^2(1 - \theta)^2(1 - \theta + \theta^2)(1 + 2\theta - 2\theta^2) + \theta^5(2 - \theta)(1 - \theta + \theta^2)$; For family 2, $L_2 \propto (1 - \theta)^6(1 - \theta + \theta^2) + \theta^4(1 - \theta)^3(2 - \theta) + \theta^3(1 - \theta)^4(1 + \theta) + \theta^6(1 - \theta + \theta^2)$.

**2.6**   Let $f_D$ = onset rate with $Dd$ or $DD$, $f_d$ = onset rate with $dd$. Then

$$p(Y11) = [\theta f_D + (1 - \theta)f_d]/2,$$
$$p(Y21) = [(1 - \theta)f_D + \theta f_d]/2,$$
$$p(N11) = [(1 - \theta)(1 - f_d) + \theta(1 - f_D)]/2,$$
$$p(N21) = [\theta(1 - f_d) + (1 - \theta)(1 - f_D)]/2.$$

$L(\theta) \propto \theta(4 + \theta)(4 - 3\theta)^2(3 - \theta)^3(6 - 5\theta)(9 - 5\theta)(9 - 7\theta)^2(8 - 7\theta)(2 + 7\theta)$
$\hat\theta = 0.20$ with $LOD = 0.47$.

**2.10**   $\hat\theta = 0.08$ with LOD score 1.486. (See $Ex210$ files on Yang's web site $AppB.3$.)

**2.11**   $\hat\theta = 0.00$ with LOD score 6.19. (See $Ex211$ files on Yang's web site $AppB.3$.)

**2.12**

| $\theta$ | mean | SD |
|---|---|---|
| 0.000 | 3.60 | 0.113 |
| 0.025 | 3.03 | 0.099 |
| 0.050 | 2.51 | 0.104 |
| 0.100 | 1.69 | 0.083 |
| 0.500 | 0.09 | 0.014 |

Details can be found in Yang's web site $AppB.4$.

**2.13**   Using Whittempre and Halpern's $\Phi_2$ score function, the z-scores for linkage are $1.323$, $-0.390$, and $0.682$ for the three families. The overall z-value is $(\sqrt{7} \times 1.323 - \sqrt{5} \times 0.390 + \sqrt{6} \times 0.682)/\sqrt{7 + 5 + 6} = 1.013$ with p-value $= 0.156$ (from the z-table) (see $Ex213$ files on Yang's web site $AppB.3$). When $V[T_j] = 0$, $T_j$ must be the same as $E[T_j]$ and this pedigree should be discard from the summation because it provides no linkage information (everyone is the same).

## Exercise 3

3.1 Hint: Use Table 2.3.2 (i) (3.1.12) is correct. (ii) (3.1.12) is correct except $E(\Delta^2|S_m = 1) = \sigma^2 + \frac{2}{5}(\Psi + 2)\sigma_a^2 + \frac{2}{5}(3\Psi^2 - 3\Psi + 1)\sigma_d^2$. However, if the children's genotypes are known, then we know $\pi_m$. For example, though $S_m = 1$ for (12) and (13), the two 1's cannot be IBD, or $\pi_m$ must be 0.
(iii) $E(\Delta^2|S_m = 0)$ non-exist,

$$E(\Delta^2|S_m = 1) = \sigma^2 + (\Psi + \tfrac{1}{2})\sigma_a^2 + \sigma_d^2.$$
$$E(\Delta^2|S_m = 2) = \sigma^2 + (\tfrac{3}{2} - \Psi)\sigma_a^2 + (2 - \Psi)\sigma_d^2.$$

3.2 (i) $Pr\{\text{Larger} = BB\} = f_{BB}^2 + 2f_{BB}f_{Bb}\Phi(\alpha_1) + 2f_{BB}f_{bb}\Phi(\alpha_1 + \alpha_2)$, where $\alpha_1 = (\mu_{BB} - \mu_{Bb})/\sigma, \alpha_2 = (\mu_{Bb} - \mu_{bb})/\sigma$.
(ii) Let $\sigma = 1$ to simplify notation.

$$Pr\{n_1 \text{ from } bb, n_2 \text{ from } Bb, n_3 \text{ from } BB\}$$
$$= \frac{n!}{n_1! n_2! n_3!} f_{bb}^{n_1} f_{Bb}^{n_2} f_{BB}^{n_3}, n_1 + n_2 + n_3 = n.$$

$$Pr\{max = BB|n_1, n_2, n_3\}$$
$$= \int_{-\infty}^{\infty} \Phi^{n_1}(x - \mu_{bb}) \Phi^{n_2}(x - \mu_{Bb}) n_3 \Phi^{n_3 - 1}(x - \mu_{BB}) \phi(x - \mu_{BB}) dx$$

The answer $Pr\{max = BB\}$ is to combine the above two formulae over all possible combinations of $n_1, n_2, n_3$ with $n_1 + n_2 + n_3 = n$.
(iii) Write down the density by conditioning on parenets and sibs' genotypes.

3.4 Hint: Let $\pi_1 = f_1 + m_1$, where $f_1 = $ IBD on the father's side for marker 1, $m_1 = $ IBD on the mother's side for marker 1, and decompose $\pi_2$ in the same way. Note $f_1$ and $m_1$ take values 0 or 1.

3.5 Hint: By (3.3.1)

3.8 $h^2 = \hat{\rho} = 0.741$. 95% C.I. = (0.61, 0.83)

3.9 (ii) $h^2 = 2\rho$ (This definition is used among geneticists, see Russell, 1990, p. 759). For $\rho = 0.5$, $h^2 = 1$ is obviously too high. The reason may be due to non-random mating, i.e., a tall person tends to marry another tall person. In this case, Cov(Father, son) $\geq \sigma_a^2/2$, or $h^2 \leq 2\rho$.

3.10 Let $Y = $ height of the father, $X = $ height of the son, $\mu = $ mean. We wish to show that

$$|E[Y|X = x] - \mu| \leq |x - \mu| \, '$$

Use (A.1.4) to show this fact.

3.11 Hint: We use mathematical induction. From $\Phi_{i-1}$ to $\Phi_i$, given $\Phi_{i-1}$ is positive definite is equivalent to $\Phi_{ii} \geq s_{i-1}' s_i$.

3.14 Both Haseman-Elston LOD and the nonparametric LOD from *MAPMAKER/SIBS* show that the gene is likely to be in the neighborhood of the second marker locus. The inputs are *Exe314.** and the outputs are *E314 * .** in *AppB.6* in Yang's web site.

## Exercise 4

4.1   $\hat{p}_A = 0.686, \hat{p}_B = 0.733, \hat{p}_{AB} = 0.512; \hat{D} = 0.01, \chi_1^2 = 0.204$, not significant.

4.2   Using $(P_{AB})_{n+1} = (P_{AB})_n + 2\theta(P_{Ab}P_{aB} - P_{AB}P_{ab})$, we have $D_{n+1} = (1 - 2\theta)D_n$. Thus, $D_n \to 0$ as $n \to \infty$.

4.3   A 95% confidence interval for $\ln(\rho)$ is $1.34 \pm 1.96 \times 0.274 = (0.804, 0.876)$. Flip it back to $\rho$.

4.4   Pick a threshold $u$ to construct a contingency table,

|  | A (Marker 1 is transmitted) | a (1 is not transmitted) |
|---|---|---|
| $B\ (Y > u)$ | $n_{AB}$ | $n_{aB}$ |
| $b\ (Y \le u)$ | $n_{Ab}$ | $n_{ab}$ |

and use the $\chi^2$ test (4.1.1), where $Y$ is a quantitative trait. To find power, we need to find $P(B|A)$ under $H_1$. Let $g_1 = (11), g_2 = (12), g_3 = (22)$, where 2 means not 1. Then

$$P(B|A) = P(B \cap A)/P(A); \quad p(A) = \sum_{i=1}^{3} P(A|g_i)p(g_i)$$

$$P(B \cap A) = \sum_{i=1}^{3} P(B|g_i)p(A|g_i)P(g_i).$$

Thus, $P(g_i)$ and $P(B)$ have to be specified, but $P(A|g_i) = 1$ if $g_i = (11); = 1/2$ if $g_i = (12)$; and 0 if $g_i = (22)$.

4.5   If the mutation rate is lower at branch $A$ than at $B$, we expect $En_{jij} > En_{ijj}$. Thus, we can let $m_1 = \sum_i \sum_{j \ne i} n_{ijj}$ and $m_2 = \sum_i \sum_{j \ne i} n_{jij}$ and use

$$Q = \frac{(m_1 - m_2)^2}{m_1 + m_2} \sim \chi_1^2$$

under $H_0$ that the mutation rates are the same to test $H_0$.

4.7   Let $H_0$ be that the $n$ mRNA's are randomly distributed on the $k$ chromosome. Then we should reject $H_0$ if too many mRNAs are concentrated on too few chromosomes. Let $x_i =$ number of genes on chromosome $i$ and $x_{(1)} \le x_{(2)} \le \cdots X_{(n)}$ are the order statistics. The reject region should be $T = \sum_{i=1}^{m} X_{(n-i+1)} \ge c$ for some threshold $c$, and the $p$-value is $Pr\{T \ge r\}$. They can be found by Monte-Carlo simulation.

4.8   The clustering steps are (1) $C + K$, $W + H$, (2) $C + K + J + V$, $B + W + H$, where $C, K$, etc. are the first letters of the races.

4.9   Proportion of silent mutation $= 22.5\%$.

4.10  The defense lawyer wants the real value as large as possible. In the example, $n = 292, \hat{f}_0 = 4.9\%$ and a $(1 - \alpha)$ upper bound for $f_0$ is

$$\sum_{i=0}^{m} \binom{n}{i} f_0^i (1 - f_0)^{n-i} \le \alpha,$$

where $m = 292 \times 4.9\% = 15$. If we let $\alpha = 0.05$, then a 95% upper confidence bound for $f_0$ is 0.08.

4.11 (i) $Pr\{\text{matching of two randomly picked alleles}\} = 2(\Sigma p_i^2)^2 - \Sigma p_i^4$.

(ii) This probability is minimized at $p_i = 1/n$ for all $i$.

(iii) For $n \geq 3$, two loci are better.

4.12 Hint: Without HWE, use the covariance formula in (A.3.3).

4.13 The $p$-values for $G_1$, $G_2$ and $G_{12}$ are 0.00045, 0.000009, and 0.001182 respectively.

# Bibliography

[1] Agresti, A. (1990) *Categorical Data Analysis.* Wiley, New York.

[2] Anderson, T. W. (1958) *Introduction to Multivariate Statistical Analysis.* Wiley, New York.

[3] Asada, Y., Varnum, D. S., Frankel, W. N. et al. (1994) A mutation in the Ter gene causing increased susceptibility to testicular teratomas maps to mouse chromosomes 18. *Nature Genetics* 6: 363–368.

[4] Ball, S. P., Cook, P. J. L., Mars M. and Buckton, K. E., (1982) Linkage between *dentinogenesis imperfacta* and GC. *Ann. Hum. Genet.* 46: 35–40.

[5] Bailey, N. T. J. (1961) *Introduction to the mathematical theory of genetic linkage.* Oxford: Clarendon Press.

[6] Beaudet, A. Browcock, A., Buchwald M. et al. (1986) Linkage of cystic fibrosis to two tightly linked DNA markers: joint report from a collaborative study. *Am. J. Hum. Genet.* 39: 681–693.

[7] Berry, D. A., Evett, I. W. and Pinch, R. (1992) Statistical inference in crime investigations using DNA profiling. *Appl. Stat.* 41: 499–531.

[8] Berry, D. A. (1991) Inferences using DNA profiling in forensic identification and paternity cases. *Statistical Science* 6: 175–205.

[9] *Biological Science: An Inquiry into Life* (1963) American Institute of Biological Sciences, Harcourt, Brace & World, Inc., New York.

[10] Blackwelder, W. C. and Elston, R. C. (1982) Power and robustness of sib-pair tests and extension to larger sibships, *Comm. Statist.—Theor. Meth.* 11: 449–484.

[11] Bodmer, W. F., Bailey, C. J., Bodmer, J. et al. (1987) Localization of the gene for familial adenomatous polynosis on chromosome 5. *Nature* 328: 614–616.

[12] Boehnke, M. (1986) Estimating the power of a proposal linkage study. *Am. J. Hum. Genet.* 39: 513–27.

[13] Breese, E. L. and K. Mather. (1957) The organisation of polygenic activity within a chromosome in *Drosophila:* I Hair characters. *Heredity* 11: 373–395.

233

[14] Breese, E. L. and Mather, K. (1960) The organisation of polygenic activity within a chromosome in *Drosophila*: II Viability. *Heredity* 14: 375–400.

[15] Buldyrev, S. V., Goldberger, A. L., Havlin, S. et al. (1993) Fractal landscapes and molecular evolution: modeling the myosin heavy chain family. *Biophysical J.* 65: 2673–2679.

[16] Cannings, C. and Thompson, E. A. (1977) Ascertainment in the sequential sampling of pedigrees. *Clin. Genet.* 12: 208–12.

[17] Cardon, L. R. and Fulker, D. W. (1994) The power interval mapping of quantitative trait loci, using selected sib pairs. *Am. J. Hum. Genet.* 55: 825–833.

[18] Carey, G. and Williamson, J. (1991) Linkage analysis of quantitative traits: increase power by using selected samples. *Am. J. Hum. Genet.* 49: 786–769.

[19] Claverie, J., Sauvaget I. and Bougueleret, L. (1990) *k*-tuple frequency analysis: From intron/exon discrimination to T-cell epitope mapping. In *Methods in Enzymology*, ed. by R. F. Doolittle. Academic Press, San Diego, 183: 237–252.

[20] Conneally, P. M., Edwards, J. H., Kidds, K. K. et al. (1985) Report of the committee on methods of linkage analysis and reporting. *Cytogenet. Cell Genet.* 40: 356–359.

[21] Cook, P. J., Robson, E. B., Buckton, K. E. et al. (1974) Segregation of genetic markers in families with chromosome variant in subject with anomalous sex differentiation. *Am. J. Hum. Genet.* 37: 261–274.

[22] Davies, J. L. et al. (1994) A genome-wide search for human type 1 diabetes susceptibility genes. *Nature* 371: 130.

[23] Devlin, B., Daniels, M. and Roeder, K. (1997) The heritability of IQ. *Nature* 338: 468–471.

[24] De Vries, R. R. P. et al. (1976) HLA-linked genetic control of host response to mycobacterium leprae. *The Lancet* 18: 1328.

[25] Doolittle, R. F. (ed.) (1990) *Molecular Evolution: Computer Analysis of Protein and Nucleic Acid Sequences*, Vol. 183 in *Methods in Enzymology*, Academic Press, New York.

[26] *DNA Technology in Forensic Science*. National Research Council. National Academic Press, 1992.

[27] Elandt-Johnson, R. C. (1971) *Probability Methods and Statistical Methods in Genetics*. Wiley.

[28] Elzo, M. A. (1996) *Animal Breeding Notes*, Animal Science Department, University of Florida.

[29] Elston, R. C. (1988) The use of polymorphic markers to detect genetic variability. *In Phenotypic variation* 88: 105.

[30] Elston, R. C. and Stewart, J. (1971) A general model for the analysis of pedigree data. *Hum. Hered.* 21: 523–42.

[31] Elston, R. C. and Keats, B. J. B. (1985) Sib-pair analyses to determine linkage groups and to order loci. *Genet. Epidemiol.* 2: 211–213.

[32] Estivill, X. et al. (1987) A candidate for the cystic fibrosis locus isolated by selection for methylation-free islands. *Nature* 326: 840–845.

[33] Finney, D. J. (1940) The detection of linkage. *Ann. Eugen.* 10: 171–214.

[34] Fisher, R. A. (1935) The detection of linkage with "dominant" abnormalities. *Ann. Eugen.* 6: 187–201.

[35] Fisher, R. A. (1936) Heterogeneity of linkage data for Friedreich's ataxia and the spontaneous antigens. *Ann. Eugen.* 7: 17–21.

[36] Freedman, D., Pisani, R. and Purves, R. (1980) *Statistics.* W.W. Norton & Co., New York.

[37] Fu, L. M. (1994) *Neural Networks in Computer Intelligence.* McGraw Hill, New York.

[38] Fulker, D. W. and Cardon, L. R. (1994) A sib-pair approach to interval mapping of quantitative trait loci. *Am. J. Hum. Genet.* 54: 1092–1103.

[39] Fulker, D. W., Cherny, S. S. and Cardon, L. R. (1995) Multipoint interval mapping of quantitative trait loci, using sib pairs. *Am. J. Hum. Genet.* 56: 1224–1233.

[40] Galton, F. (1889) *Natural Genetics.* Macmillan, London.

[41] Gardner, E. J., Simmons, M. J. and Snustad, D. P. (1991) *Principles of Genetics.* Eighth Edition, Wiley. (The 1981 edition was by Gardner and Snustad.)

[42] Gasser, C. S. and Fraley R. T. (1992) Transgenic crops. *Scientic American* June: 62–69.

[43] Goddard, M. E. (1992) A mixed model for analyses of data on multiple genetic markers. *Theor. Appl. Genet.* 83: 878–886.

[44] Gordon, J. W., Scangos, G. A., Plotkin, D. J., Barbosa, J. A. and Ruddle, R. H. (1980) Genetic transformation of mouse embryos by microinjection of purified DNA. *Proc Natl Acad. Sci. USA* 77: 7380–84.

[45] Green, J. R., Low, H. C. and Woodrow, J. C. (1983) Inference of inheritance of disease using reptition of HLA haplotypes in affected siblings. *Ann. Hum. Genet.* 47: 73–82.

[46] Green, J. R. and Low, H. C. (1984) The distribution of $N$ and $F$: Measures of HLA haplotype concordance. *Biometrics* 40: 341–348.

[47] Green, J. R . and Grennan, D. M. (1991) Testing for haplotype sharing by siblings with incomplete information of parental haplotypes. *Ann. Hum. Genet.* 55: 243–249.

[48] Green, J. R. and Shah, A. (1992) The power of the N-test haplotype concordance. *Ann. Hum. Genet.* 56: 331–338.

[49] Gusella, J. F., Wexler, N. C., Conneally, P. M. et al. (1983) A polymorphic marker genetically linked to Huntington's disease. *Nature* 306: 234–238.

[50] Haldane, J. B. S. and Smith, C. A. B. (1947) A new estimate of the linkage between the genes for colour-blindness and haemophilia in man. *Ann. Eugen.* 14: 10–31.

[51] Haley, C. S. and Knott, S. A. (1992) A simple regression method for mapping quantitative trait loci in line crosses using flanking markers. *Heredity*, 69: 315–324.

[52] Harris, A., Shackleton, S. and Hull, J. (1996) Mutation detection in the cystic fibrosis transmembrane conductance regulator gene. *Methods in Molecular Genetics* (ed. by K. W. Adolph) 8: 70–96.

[53] Hartmann, J. et al. (1994) Diversity of ethnic and racial VNTR RFLP fixed-bin frequency distribution. *Am. J. Hum. Genet.* 55: 1268–78.

[54] Haseman, J. K. and Elston, R. C. (1972) The investigation of linkage between a quantitative trait and a marker locus. *Behavior genetics* 2: 3–19.

[55] Henderson, C. R. (1976) A simple method for computing the inverse of a numerator relationship matrix using in prediction of breeding values. *Biometrics* 32: 69–83.

[56] Hill, W. G. and Robertson, A. (1968) Linkage disequilibrium in finite populations. *Theor. Appl. Genet.* 38: 226–231.

[57] Hill, W. G. and Weir, B. S. (1994) Maximum-likelihood estimation of gene location by linkage disequilibrium, *Am. J. Hum. Genet.* 54: 705–714.

[58] Holmans, P. (1993) Asymptotic properties of affected-sib-pair linkage analysis. *Am. J. Hum. Genet.* 52: 362–374.

[59] Hogben, L. and Pollack, R. (1935) A contribution to the relation of the gene loci involved in the isoagglutinin, taste blindnes, Friedreich's ataxia, and major brachydactyly of man. *J. Genet.* 31: 353–361.

[60] Hughes, A. L. and Yeager, M. (1997) Comparative evolutionary rates of introns and exons in murine rodents. *J. Mol. Evol.* 45: 125–130.

[61] Jacquard, A. (1974) *The Genetic Structure of Population.* Springer-Verlag, New York.

[62] Jiang, C. and Zeng, Z.-B. (1995) Multiple trait analysis of genetic mapping for quantitative trait loci. *Genetics* 140: 1111–1127.

[63] Jorde, L. B. et al. (1994) Linkage disequilibrium predicts physical distance in the Adenomatous polyposis coli region. *Am. J. Hum. Genet.* 54: 884–898.

[64] Johnson, N. L. and Kotz, S. (1969) *Discrete Distributions.* John Wiley, New York.

[65] Jukes, T. H. and Cantor, C. R. (1969) Evolution of protein molecules, in H. N. Munro (ed.), *Mammalian Protein Metabolism.*, pp. 21–132. Academic Press, New York.

[66] Julier, C. et al. (1991) Insulin-IGF2 region on chromosome 11p encodes a gene implicated in HLA-DR-4-dependent diabetes susceptibility. *Nature* 354: 155–159.

[67] Kaplan, N. L., Martin, E. R. and Weir, B. S. (1997) Power studies for the transmission/disequilibrium tests with multiple alleles. *Am. J. Hum. Genet.* 60: 691–702.

[68] Karlin, S. (1992) R. A. Fisher and evolutionary theory. *Statistical Science* 7: 13–33.

[69] Keim, P., Driers, B. W., Olson, T. C. et al. (1990) RFLP mapping in soybean: Association between marker loci and variation in quantitative traits. *Genetics* 126: 735–42.

[70] Kimura, M. (1980) A simple method for estimating evolutionary rate of base substitutions through comparative studies of nucleotide sequences. *J. Mol. Evol.* 17: 111–20.

[71] Kimura, M. and Crow, J. F. (1964) The number of alleles that can be maintained in finite population, *Genetics* 49: 725–738.

[72] King, R. C. and Stansfield, W. D. (1990) *A Dictionary of Genetics*, Oxford University Press, New York.

[73] Kingsman, S. M. and Kingsman, A. J. (1988) *Genetics Engineering.* Blackwell Scientific Publication, Boston.

[74] Klug, W. S. and Cummings, M. R. (1983) *Concepts of Genetics.* Charles E. Merril Pub. Co., Ohio.

[75] Kruglyak, L. and Lander, E. S. (1995) Complete multipoint sib-pair analysis of qualitative and quantitative traits. *Am. J. Hum. Genet.* 57: 439–454.

[76] Lanave, C., Preparata, G., Saccone, C. and Serio, G. (1984) A new method for calculating evolutionary substitution rates, *J. Mol. Evol.* 20: 86–93.

[77] Lande, R. and Thompson, R. (1990) Efficiency of Marker-assisted selection in the improvement of quantitative traits. *Genetics* 124: 743–756.

[78] Lander, E. S. and Botstein, D. (1989) Mapping Mendelian factors underlying quantitative traits using RFLP linkage maps. *Genetics* 121: 185–199.

[79] Lander, E. S. and Budowle, B. (1994) DNA fingerprinting dispute laid to rest. *Nature* 371: 735.

[80] Lander, E. S. and Schork, N. J. (1994) Genetic dissection of complex traits. *Science* 265: 2037–47.

[81] Lange, K. (1986) A test statistic for the affected-sib-set method. *Ann. Hum. Genet.* 50: 283–290.

[82] Lange, K. (1997) *Mathematical and Statistical Methods for Genetic Analysis.* Springer-Verlag, New York.

[83] Larget-Piet, D., Gerber, S., Bonneau, D. et al. (1994) Genetic heterogeneity of usher syndrome type 1 in French families. *Genomics* 21: 138–143.

[84] Lathrop, G. M., Lalouel, J. M., Julier, C. and Ott, J. (1985) Multilocus linkage analysis in humans: Detection of linkage and estimation of recombination. *Am. J. Hum. Genet.* 37: 482–98.

[85] Liberman, U. and Karlin, S. (1984) Theoretical models of genetic map functions. *Theoretical Population Biology* 25: 331–346.

[86] Littell, R. C., Milliken, G. A., Stroup, W. W. and Wolfinger, R. D. (1996) *SAS System for Mixed Models,* SAS Institute Inc., Cary, N.C.

[87] Lucassen, A. M., Julier, C., Beressi, J. (1993) Susceptibility to insulin dependent diabetes mellitus maps to a 4.1 kb segment of DNA spanning the insulin gene and associated VNTR. *Nature Genetics* 4: 305–309.

[88] Luxmoore, R. A. (ed.) (1992) *Directory of Crocodilian Farming Operations,* IUCN-The World Conservation Union, UK.

[89] Marron M. P. et al. (1997) Insulin-dependent diabetes mellitus (IDDM) is associated with CTLA4 polymorphisms in multiple ethnic groups. *Human Molecular Genet.* 6: 1275–1282.

[90] Marshall, E. (1995) Gene therapy's growing pains. *Science* 269: 1050–1055.

[91] Martin, E. R., Kaplan, N. L. and Weir, B. S. (1997) Tests for linkage and association in nuclear families. *Am. J. Hum. Genet.* 61: 439–448.

[92] Martinez, O. and Curnow, R. N. (1992) Estimating the locations and sizes of the effects of quantitative trait loci using flanking markers. *Theor. Appl. Genet.* 85: 480–488.

[93] Mather, K. and Jinks, J. L. (1971) *Biometrical Genetics: The Study of Continuous Variation.* Cornell University Press, Ithaca, New York.

[94] Mather, K. (1941) Variation and selection of polygenic characters. *J. Genet.* 41: 159–193.

[95] Mather, K. (1942) The balance of polygenic combinations. *J. Genet.* 43: 309–336.

[96] Meng, X.-L. and Rubin, D. B. (1993) Maximum likelihood estimation via the ECM algorithm: a general framework. *Biometrika* 80: 276–268.

[97] Mendel, G. (1867) Letter to Carl Nägeli. English translation in *Great Experiments in Biology,* edited by M. G. Gabriel and S. Fogel, Prentice -Hall, 1955.

[98] Montgomery, D. C. (1984) *Design and Analysis of Experiments.* Wiley, New York.

[99] Morton, N. E. (1955) Sequential tests for the detection of linkage. *Am. J. Hum. Genet.* 7: 277–318.

[100] Morton, N. E. (1956) The detection and estimation of linkage between the genes for elliptocytosis and the Rh blood type. *Am. J. Hum. Genet.* 8: 80–96.

[101] Morton, N. E. (1991) Parameters of the human genome. *Proc. Natl. Acad. Sci. USA* 88: 7474–7476.

[102] Nei, M. (1987) *Molecular Evolutionary Genetics.* Columbia University Press, New York.

[103] Ott, J. (1974) Estimation of the recombination fraction in human pedigrees: Efficient computation of the likelihood for human linkage studies. *Am. J. Hum. Genet.* 26: 588–597.

[104] Ott, J. (1991) *Analysis of Human Genetic Linkage,* Revised ed. Johns Hopkins University Press, Baltimore.

[105] Pascoe, L. L. and Morton, N. E. (1987) The use of map functions in multipoint mapping. *Am. J. Hum. Genet.* 40: 174–83.

[106] Paterson, A. H. et al. (1990) Fine mapping of quantitative trait loci using selected overlapping recombinant chromosomes, in an interspecies cross of tomato. *Genetics* 124: 735–742.

[107] Paterson, A. H., Lander, E. S., Hewitt, J. D. et al. (1988) Resolution of quantitative traits into Mendelian factors by using a complete linkage map of restriction fragment length polymorphisms. *Nature* 335: 721–726.

[108] Pennisi, E. (1994) Fanfare over finding first gene. *Science News* December 3 issue: 372.

[109] Pennisi, E. (1997) Transgenic lambs from cloning lab. *Science* December 22: 631.

[110] Penrose, L. S. (1935) The detection of autosomal linkage in data which consist of pairs of brothers and sisters of unspecified parentage. *Ann. Eugenics* 6: 133–8.

[111] Ploughman L. M. and Boehnke, M. (1989) Estimating the power of a proposed linkage study for a complex genetic trait. *Am. J. Hum. Genet.* 44: 543–551.

[112] Pras, E., Arber, A., Aksentijevich, I. et al. (1994) Localization of a gene causing cystinuria to chromosome 2p. *Nature genetics* 6: 415.

[113] Quaas, R. L. (1988) Additive genetic model with group and relationships. *J. Dairy Sci.* 71: 1338–45.

[114] Renwick, J. H. and Bolling, D. R. (1967) A program complex for encoding, analyzing and storing human linkage data. *Am. J. Hum. Genet.* 19: 360–67.

[115] Renwick, J. H. and Schulze, J. (1965) Male and female recombination fraction for the nail-patella: ABO lonkage in man. *Am. J. Hum. Genet.* 28: 379–92.

[116] Richards, F. M. (1991) The protein folding problem. *Scientific American*, January Issue: 54–63.

[117] Risch, N. (1987) Assessing the role of HLA-linked and unlinked determinants of disease. *Am. J. Hum. Genet.* 40: 1–14.

[118] Risch, N. (1990a) Linkage strategies for genetically complex traits I. Multilocus models. *Am. J. Hum. Genet.* 46: 222–228.

[119] Risch, N. (1990b) Linkage strategies for genetically complex traits II. The power of affected relative pairs. *Am. J. Hum. Genet.* 46: 229–241.

[120] Risch, N. (1990c) Linkage strategies for genetically complex traits III. The effect of marker polymorphism on analysis of affected relative pairs. *Am. J. Hum. Genet.* 46: 242–253.

[121] Rise, M. L., Frankel, W. N., Coffin, J. M. et al. (1991) Genes for epilepsy mapped in the mouse. *Science* 253: 669–673.

[122] Roeder, K. (1994) DNA Fingerprinting: A review of the controversy. *Statistical Science* 9: 222–278.

[123] Roff, D. A. (1997) *Evolutionary Quantitative Genetics*. Chapman & Hall, New York.

[124] Rowen, L., Mahariras, G. and Hood, L. 1997 Sequencing the human genome. *Science* 278: 605–607.

[125] Rouleau, G. A., Wertelecki, W., Haines, J. et al. (1987) Genetic linkage of bilateral acoustic neurofibromatosis to a DNA marker on chromosome 22. *Nature* 329: 246–248.

[126] Russell, P. J. (1990) *Genetics*, Second Edition, Scott, Foreman and Co. Glenview, Ill.

[127] Saiki, R. K., Bugawan, T. L. et al. (1986) Analysis of enzymatically amplified β-globin and HLA-DQ DNA with allele-specific oligonucleotide probes. *Nature* 324: 163–166.

[128] Sampson A. R. (1974) A tale of two regressions, *J. Amer. Statist. Assoc.* 69: 682–689.

[129] Sapienza, C. (1990) Parental imprinting of genes. *Scientific American*, October issue, 52–60.

[130] Scarr, S. and Carter-Saltzman, L. (1983) Genetics and intelligence (1983) In *Behavior Genetics: Principles and Applications*, ed. by J. L. Fuller and E. C. Simmel, Lawrence Erlbaum Assoc., New Jersey.

[131] Schork, N. J. and Schork, M. A. (1993) The relative efficiency and power of small pedigree studies of the heritability of a quantitative trait, *Hum. Hered.* 43: 1–11.

[132] Searle, S. R., Casella, G. and McCulloch, C. E. (1992) *Variance Components.* Wiley, New York.

[133] Self, S. G. and Liang, K. Y. (1987) Asymptotic properties of maximum likelihood estimators and likelihood ratio tests under nonstandard conditions. *J. Amer. Statist. Assoc.* 82: 605–610.

[134] Senapathy, P., Shapiro, M. B. and Harris, N. L. (1990) Splice junction, branch point sites, and exons: sequence statistics, identification, and applications to geneome project. In *Methods in Enzymology*, ed. by R. F. Doolittle, Academic Press, San Diego, 183: 252–278.

[135] Sham, P. C. and Curtis, C. (1995) An extended transmission/disequilibrium test (TDT) for multi-allele marker loci. *Ann. Hum. Genet.* 59: 323–336.

[136] Smith, C. A. B. (1961) Homogeneity test for linkage data. *Proc. Sec. Int. Congr. Hum. Genet.* 1: 212–13.

[137] Smith, C. A. B. (1963) Testing for heterogeneity of recombination fraction values in human genetics. *Ann. Hum. Genet.* 27: 175–82.

[138] Solovyev, V. V., Salamov, A. A. and Lawrence, C. B. (1994) Predicting internal exons by oligonucleotide composition and discriminant analysis of spliceable open reading frames. *Nucleic Acids Research* 22(24): 5156–5163.

[139] Spence, J. E., Rosenbloom, C. L., O'Brein, W. E. et al. (1986) Linkage of DNA markers to cystic fibrosis in 26 families. *Am. J. Hum. Genet.* 39: 729–734.

[140] Spieces, E. B. (1977) *Genes in Population.* Wiley, New York.

[141] Spielman, R. S., McGinnis, R. E. and Ewens, W. J. (1993) Transmission test for linkage disequilibrium: The insulin gene region and insulin-dependent diabetes mellitus (IDDM). *Am. J. Hum. Genet.* 52: 506–516.

[142] Staden, R. (1990) Finding protein coding regions in genomic sequences. In *Method in Enzymology* ed. by R. F. Doolittle. Academic Press, San Diego, 183: 163–180.

[143] Stuart, A. and Ord, J. K. (1987) *Kendall's Advanced Theory of Statistics, Vols. I and II*. Oxford Univ. Press, New York.

[144] Stuart, A. and Ord, J. K. (1991) *Kendall's Advanced Theory of Statistics, Vol. II*. Oxford Univ. Press, New York.

[145] Suzuki, D. T., Griffiths, A. J. F. and Lewontin, R. C. (1981) *An Introduction to Genetic Analysis*, W. H. Freeman and Company, San Francisco.

[146] Sved, J. A. (1971) Linkage disequilibrium and homozygosity of chromosome segments in finite populations. *Theo. Popul. Bio.* 2: 125–141.

[147] Tajima, F. (1993) Simple methods for testing the molecular evolutionary clock hypothesis. *Genetics* 135: 559–607.

[148] Tchernev, V. T., Maria, D. F., Barbosa, F. S. et al. (1997) Genetic mapping of 20 novel expressed sequence tags from midgestation mouse embryos suggests chromsomal clustering. *Genomics* 40: 170–174.

[149] Tellegen, J. A., Lykken, D. T., Bouchard, T. J. et al. (1988) Personality similarity in twins reared apart and together. *J. Personality and Social Psych.* 54: 1031–1039.

[150] Terwilliger, J. D. (1995) Assessing the role of HLA-linked and unlinked determinants of disease. *Am. J. Hum. Genet.* 56: 777–787.

[151] Terwilliger, J. D. and Ott, J. (1994) *Handbook of Human Genetic Linkage*, The Johns Hopkins University Press, Baltimore.

[152] Thompson, E. A. (1986) *Pedigree analysis in human genetics*. Baltimore: Johnson Hopkins University Press.

[153] Thompson, M. W., McInnes, R. R. and Willard, H. F. (1991) *Genetics in Medicine*, 5th Edition, W. B. Saunders Co. Philadelphia.

[154] Thomson, G. (1980) A two locus model for juvenile diabetes. *Ann. Hum. Genet.* 43: 383–398.

[155] Vaeck, M., Reynaerts, A., Hofte, H. et al. (1987) Transgentic plants protected from insect attack. *Nature* 328: 33–37.

[156] Van Arendonk, J. A. M., Tier, B. and Kinghorn, B. P. (1994) Use of multiple Genetic markers in prediction of breeding values. *Genetics* 137: 319–329.

[157] Wald, A. (1947) *Sequential Analysis*. Wiley, New York. Reprinted by Dover Publications Inc. (1973).

[158] Warren, D. C. (1924) Inheritance of eqq size in *Drosophila melanogaster*. *Genetics* 9: 41–69.

[159] Watkins, P. C., Schwartz, R. and Hoffman, N. et al. (1986) A linkage study of cystic fibrosis in extended multigenerational pedigrees. *Am. J. Hum. Genet.* 39: 735–743.

[160] Weaver, R. F. and Hedrick, P. W. (1989) *Genetics*. WM. C. Brown Publishers, Iowa.

[161] Weitkamp, L. R. et al. (1981) Repressive disorder and HLA: a gene on chromosome 6 that can affect behavior. *New Eng. J. Med.* 305: 1301–1306.

[162] Weir, B. S. (1996) *Genetic Data Analysis II*, Sinauer Assoc. Sunderland, Mass.

[163] Weir, B. S. (1992) Population genetics in the forensic DNA debate, *Proc. Natl. Acad. Sci.* 89: 11654–11659.

[164] White, H. (1989) Some asymptotic results for learning in single hidden-layer feedforward network models, *J. Am. Statist. Assoc.* 86: 1003–1013.

[165] Whittemore, A. S. and Halpern, J. (1994a) A class of tests for linkage using affected pedigree members. *Biometrics* 50: 118–127.

[166] Whittemore, A. S. and Halpern, J. (1994b) Probability of gene identity by descent: Computation and applications. *Biometrics* 50: 109–117.

[167] Xu, S. (1995) A comment on the simple regression method for interval mapping. *Genetics* 141: 1657–1659.

[168] Xu, Y., Muarl, R., Shah, M. and Uberbacher, E. (1994) Recognizing exons in genomic sequence using GRAIL II. *Genetic Engineering* 16: 241–253.

[169] Yang, M. C. K. (1995a) A survey of statistical methods in modern genetics. *J. Chin. Stat. Asso.* 33: 407–428.

[170] Yang, M. C. K. (1995b) Heredity, environment and gene. *Mathmedia* 19: 14–21.

[171] Yang, Z. (1994) Estimating the pattern of nucleotide substitution, *J. Mol. Evol.* 39: 105–111.

[172] Zhang, W. and Smith, C. (1992) Computer simulation of marker-assisted selection utilizing linkage disequilibrium. *Theor. Appl. Genet.* 83: 813–820.

[173] Zeng, Z.-B. (1993) Theoretical basis for separation of multiple linked gene effects in mapping quantitative trait loci. *Proc. Natl. Acad. Sci. USA* 90: 10972–10976.

[174] Zeng, Z.-B. (1994) Precision mapping of quantitative trait loci. *Genetics* 136: 1457–1468.

[175] Zhao, L. P., Thompson, E. and Prentice, R. (1990) Joint estimation of recombination fractions and interference coefficients in multilocus linkage analysis. *Am. J. Hum. Genet.* 47: 255–65.

[176] Zieve, I., Wiener A. S., and Fries, J. H. (1936) On the linkage relations of the genes for allergic disease, and the gene determining the blood group, *MN* types and eye color in man. *Ann. Eugen., London* 7: 163–178.

# Index